Frank Swift Bourns, Dean C. Worcester

Preliminary Notes on the Birds and Mammals collected by

the Menage Scientific Expedition to the Philippine Islands

Frank Swift Bourns, Dean C. Worcester

Preliminary Notes on the Birds and Mammals collected by the Menage Scientific Expedition to the Philippine Islands

ISBN/EAN: 9783744716468

Printed in Europe, USA, Canada, Australia, Japan

Cover: Foto ©berggeist007 / pixelio.de

More available books at **www.hansebooks.com**

THE MINNESOTA ACADEMY OF NATURAL SCIENCES
At Minneapolis, Minn.
OCCASIONAL PAPERS, VOL. I., NO. 1

PRELIMINARY NOTES

ON THE

BIRDS AND MAMMALS

COLLECTED BY THE

MENAGE SCIENTIFIC EXPEDITION

TO THE

·PHILIPPINE ISLANDS

By Frank S. Bourns and Dean C. Worcester

MINNEAPOLIS
PRINTED FOR THE ACADEMY
December 1894

EDITORIAL NOTE.

With this brochure the Trustees of the Minnesota Academy of Natural Sciences enter upon the issue of a new series of publications. It is termed "Occasional Papers." The series is instituted to enable the Academy to print the researches of its members when such researches attain the measure of a monograph or memoir. It will also enable the Trustees to prepare and distribute the results of investigations more promptly than could be done were such material to await its place in the series of Bulletins. It is hoped that the reasons which have prevented the inception of the series of Occasional Papers will not in the future prevent the issue of such material as may be deemed worthy of publication.

The specimens described in the following "Preliminary Notes" as well as the entire collection of the Menage Scientific Expedition to the Philippine Islands, are the property of the Minnesota Academy of Natural Sciences. An exchange list will be issued to which the attention of all zoologists and of ornithologists in particular will be invited.   C. W. HALL.

PRELIMINARY NOTES ON THE BIRDS AND MAMMALS COLLECTED BY THE MENAGE SCIENTIFIC EXPEDITION TO THE PHILIPPINE ISLANDS.

BY FRANK S. BOURNS AND DEAN C. WORCESTER.

I.

INTRODUCTION.

The writers of the present paper had the honor of forming two of the "party of five collectors from the United States" which consti tuted the Steere Expedition to the Philippines. In company with Dr. Steere they visited, in 1887–88, thirteen of the larger islands of the group. The birds and mammals collected by them were placed at the disposal of Dr. Steere for identification and description.

Being convinced that much remained to be done, both in the discovery of new species and in the working out of the exact distribution of species already known, we were extremely anxious to return and continue the work. This we were enabled to do in the summer of 1890 through the liberality of Mr. Louis F. Menage a public spirited citizen of Minneapolis, Minnesota, and a member of the Minnesota Academy of Natural Sciences.

The entire expense of the expedition was borne by Mr. Menage and its results were donated to the Academy of Sciences.

We found in Mr. Menage a supporter careful as to the ends for which the funds he supplied were expended, but quick to see and appreciate the needs of the expedition and always ready to meet them.

We originally planned to spend two years in the field. As the time allowed us drew near its end, and it became evident that we could not complete the work we desired to do before its expiration, we asked for an extension of time which Mr. Menage readily granted. At the end of two years we separated, Mr. Bourns going to Borneo to collect the interesting mammals of that region and Mr. Worcester remaining in the Philippines to complete the work there.

When all the localities in which we had planned to make collections, with the exception of North Luzon and the Babuyanes and Batanes islands, had been visited a serious attack of illness brought Mr. Worcester's work to an end and this promising field, from which we had anticipated much, had to be left unexplored.

During the stay of two years and five months the following islands were visited: Luzon, Samar, Mindanao, Basilan, Sulu, Tawi Tawi, Palawan Culion, Busuanga, Mindoro, Tablas, Romblon, Sibuyan, Panay, Guimaras, Negros, Cebu, Siquijor.

At the time of our visit Tawi Tawi, Tablas, Romblon and Sibuyan were new localities, and though we learned from the natives that the French Naturalist, Marche, had visited the Calamianes Islands (Culion and Busuanga) we have been unable to find any record of his collections.

We returned to this country in the early summer of 1893 expecting immediately to begin work on our material, but the financial troubles prevailing at that time had seriously embarrassed many friends of the Academy, so that during the stagnation that followed the panic of 1893 the work came to a standstill and could not be resumed until the summer of the present year.

Meanwhile Mr. A. Everett, the well known Philippine and Bornean collector, had visited Sibutu and Bongao and sent his collectors to Tawi Tawi, so that several of our most important discoveries in the latter island have been anticipated.

The following preliminary notes will be followed at a later date by a more extended paper in which many results of our trip which can not here be even mentioned will be brought out and their exact bearing on the work of our predecessors made plain.

We wish to express our very sincere gratitude to Dr. Thomas S. Roberts, Professor Henry F. Nachtrieb, Professor Henry L. Osborn,

Mr. Edward C. Gale and Mr. Horace V. Winchell for the large amount of time which they have given to the carrying out of our common plans; to the Athenaeum Library Board and the Minneapolis Public Library for their liberality in providing the literature without which the work would have been impossible and to the many friends of the Academy who have furnished financial aid at a time when they could ill afford to do so.

In connection with our final paper we shall publish careful measurements of all the species of land birds obtained. With the exception of the birds collected by Mr. Moseley no measurements were taken in the field on the birds of the Steere Expedition and many of the length measurements given for them were taken from dry skins and are, therefore, unreliable. With very few exceptions all of the birds collected by the Menage Expedition were measured *in the flesh* for length and during the past summer a complete and careful series of measurements of more than four thousand specimens has been prepared by Messrs. Lawrence E. Griffin and Ernest G. Martin of Hamline University, Saint Paul, Minn. We are greatly indebted to both of these young men for the care and diligence with which they carried out this important piece of work. The data furnished by them will enable us not only to furnish for each species average measurements from a large series of specimens but also to ascertain the relative amount of individual variation in the representatives of those genera which display a strong tendency to develop local species as compared with other genera which show no such tendency. In addition to the work above mentioned Messrs. Griffin and Martin gave us much valuable assistance.

II.

NEW SPECIES OF BIRDS.

The following species are believed to be new:

1. **Ninox spilonotus** *sp. nov.*

Sexes alike. General color of upper surface fulvous brown. Feathers of head, nape, interscapulars and wing-coverts spotted with light rufous brown giving the parts in question a decidedly speckled appearance. Rump fulvous brown, upper tail-coverts faintly spotted with pale rufous brown. Tail nearly black with nine narrow transverse bands of light rufous brown. Quills like tail but spotted, instead of barred, with light rufous brown. Scapulars like back, some of them with large nearly white spots on outer webs. A few of greater series of wing-coverts also spotted with white on outer webs. Chin and throat whitish, almost pure white in one specimen, in others light rufous brown, the feathers always with black shaft stripes. Auriculars fulvous brown somewhat mixed with light rufous brown. Sides of neck like head. Breast, abdomen, flanks and thighs, under wing coverts and axillaries rufous brown, the depth of the color subject to great individual variation. Many feathers of breast and abdomen with fulvous brown spots and all with blackish bases. Under surface of wing fulvous brown. Inner webs of feathers, especially of secondaries, spotted and barred with light rufous brown. A spot of white on bend of wing. Tarsus feathered for rather more than half its length. Iris yellow, legs and feet pale yellow, bill black at tip, gray at base. Two females measure 9.50 inches in length. Culmen, .53. Tarsus, 1.21. Wing, 7.13. Tail, 4.00.

Habitat: Cebu, Sibuyan, Tablas, Mindoro.

A single specimen of this species was secured in Cebu by Mr. Worcester in 1888. Its curious mottled back and under surface were suggestive of immaturity, and Dr. Steere thought it to be the young of some undescribed species. We have altogether too much material now to make such a theory tenable.

2. **Phabotreron cinereiceps** *sp. nov.*

Top of head, nape and sides of neck clear ashy gray, slightly washed with rufous on forehead. Hind-neck amethystine as in P.

amethystina. Back, rump and upper tail-coverts brown with bronze reflections, the tail coverts slightly more ruddy than back. Four outer pairs of tail feathers dark brown, lighter at base. Two central pairs ruddy brown with bronze reflections. All the tail feathers with ashy tips which form a distinct terminal band ¼ inch in width. Shafts of tail feathers black. Wing coverts and secondaries uniform with back. Primaries dark brown, the first five sharply edged with rusty brown on outer web. A narrow black stripe under eye. Sides of face, ear-coverts, fore-neck and breast rich ruddy brown, the breast with slight metallic gloss. Chin and throat lighter. Abdomen and thighs fulvous brown. Flanks darker with slight metallic wash. Under tail-coverts clear ashy gray. Shafts of tail feathers with basal half black, apical half white. Under surface of tail nearly black, the terminal grey band distinct and wider than on upper surface, measuring .6 inch on outer pair of feathers. Under wing-coverts and axillaries like the flanks. Under surface of quills uniform dark brown. Bill black. Legs and feet dirty purplish. Nails black. Iris in one specimen bright yellow, in another orange red. Length, 10.25 in. Culmen, .80. Wing, 5.29. Tail, 3.90. Tarsus, .70. Sexes alike.

Habitat: Tawi Tawi.

3. Phabotreron brunneiceps *sp. nov.*

Above dark brown with greenish reflections. Amethystine spot on hind neck less blue than P. amethystina. Top of head brown, the forehead slightly lighter and nape slightly darker than crown. Narrow dark brown streak under eye. Sides of face and ear-coverts brown, paler than crown. Narrow dark brown streak under eye. Chin and throat grayish fulvous. Breast pearly ash. Abdomen, flanks, thighs and under tail-coverts ochraceous brown. Under surface of tail brownish black, with broad grey terminal band. Under wing-coverts and axillaries fulvous brown. Primaries with sharply defined light edges on upper web. Below slightly more ashy. Tail-feathers brown above with distinct terminal bands of grey. Central pair with slight metallic gloss. Shafts of quills black above and below except the terminal half inch which is white. Bill black. Feet dark pink. Nails brown. Iris orange red. Length, 10.33 inches. Culmen, .94. Wing, 5.21. Tail, 3.50. Tarsus, .72. A well defined species readily distinguished from P. amethystina by its smaller size and the entirely different color of its under surface.

Habitat: Basilan.

4. Phabotreron maculipectus *sp. nov.*

Adult male: Upper surface exactly as in P. amethystina except that the primaries are slightly darker. Dark brown stripe under eye extending from gape through ear-coverts to hind-neck. Below this a white stripe and a second shorter dark stripe below the latter. Cheeks fulvous brown. Chin and throat more ruddy brown. Breast clear ashy grey, each feather having an edging distinctly lighter than its center, producing a beautiful mottled appearance. Feathers on center of fore-breast washed with brown and forming a distinct patch. Feathers of abdomen lack the dark centers, and their edges are washed with light brown. Thighs and under tail-coverts cinnamon brown, much lighter than in P. amethystina. Under surface of tail-feathers dark brown, nearly black, with faint metallic gloss and a broad grey terminal band. Shafts of feathers black changing to white at tips. Under surface of wing and axillaries uniform fulvous brown. Bill black. Feet dark pink. Nails dark brown, nearly black. Culmen, 1.02. Wing, 5.69. Tail, 4.55. Tarsus, .77. Length not taken from birds in flesh. This beautiful species was obtained in the island of Negros on the mountains of the interior, where it is by no means common. It is distinguished from all the other species of the genus by its fine mottled breast.

5. Phabotreron frontalis *sp. nov.*

General color of upper surface as in P. brunneiceps but forehead and crown lighter, nape washed with ashy grey, and lacking metallic gloss. Tail glossed with dull bronze instead of amethystine and terminal band less strongly marked than in brunneiceps. Under surface much as in brunneiceps but everywhere darker. Under tail coverts ashy grey slightly tipped with fulvous. Tail much as in brunneiceps, the outer web of outer pair of feathers being, however, light brown. Basal half of shafts dirty whitish. Apical fourth white, rest brown. Iris pale orange. Bill black. Legs and feet purple. Nails light brown. Sexes alike. Length, 10.37 inches. Culmen, 2.00. Wing, 5.57. Tail, 3.95. Tarsus, .81.

Habitat: Cebu.

6. Phlogoenas menagei *sp. nov.*

Entire upper surface of head, nape. hind-neck, upper back, sides of neck and sides of breast rich metallic green. Scapulars and interscapulars dark brown, broadly edged with elegant violet when specimen is held between observer and the light, this color changing to deep green when specimen is held away from source of light. Rump and upper tail-coverts ruddy brown narrowly edged with the metallic colors of the back. A few of the longest coverts nearly black, washed with rufous brown at the tips. Basal portion of tail feathers dark ashy grey, the two central feathers darkest. A distinct sub-terminal band of black on all but the central pair of feathers. All the feathers with a terminal grey band, least distinct on central pair. Wing coverts dark brown, broadly tipped with metallic green except outer series, which are broadly tipped with ashy grey. Primary and secondary coverts and secondaries fulvous brown, the outer half of outer webs of feathers rich rufous brown, the inner secondaries having the entire outer web, and tip of inner web, of this color. Primaries fulvous brown faintly washed with rufous brown on basal half of outer webs. Lores, a narrow line under eye and ear-coverts nearly black with a faint wash of metallic green. The metallic green of back and sides of neck continued in a distinct band across the breast, only slightly interrupted at center of breast and enclosing a beautiful orange plastron formed by the bristle-like tips of the feathers of the fore breast. Basal portion of these feathers as well as chin, throat, sides of face and sides of throat pure white. An indistinct white band behind the green pectoral band. Hind breast and upper abdomen pearly ash, a few of the feathers tipped with creamy white. Belly creamy white. Flanks, thighs and under tail-coverts buff. Under surface of tail like upper, the terminal band being rather more pronounced. Under wing coverts, axillaries and basal portion of inner webs of all the quills chestnut brown. Rest of quills dark brown. Bill slaty grey at tip, black at base. Legs and feet light red. Nails light brown. Iris light silver grey. Length, 11.25 inches. Culmen, .85. Wing, 6.03. Tail, 4.07. Tarsus, 1.43.

Extremely rare and difficult to obtain. We secured two fine males but failed to get a female.

Habitat: Tawi Tawi.

We take pleasure in naming this fine bird in honor of Louis F. Menage, through whose liberality our second visit to the Philippine Islands was made possible.

7. Batrachostomus menagei sp. nov.

Adult male: Top of head rich dark brown slightly washed with black. Feathers of forehead buff, tipped with fulvous brown, forming a distinct buff stripe reaching back to eye. Feathers of crown lighter fulvous with spots of rufous brown on the edges, each spot being surrounded with black. Some of the feathers tipped with rufous, and having black sub-terminal bands. Occiput and nape with less black. Elongated auriculars tawny buff, with black spots and bars, the tips being black. Sides of face tawny buff streaked with black, lighter below. A distinct buffy white nuchal collar formed by white subterminal bars on feathers of neck, the bases of which are dark buff thickly vermiculated with black. Their tips are black, and a black band intervenes between the white subterminal band and the buffy bases of the feathers. Feathers of back dark brown, thickly vermiculated with black. Feathers of rump fulvous brown, spotted with black and reddish brown toward their tips, these colors assuming the form of irregular bands on the upper tail-coverts A few of the shorter scapulars almost black with irregular bars of dark rufous brown. Outer webs of longer scapulars light buff, the two outermost feathers being entirely of this color. The next scapulars have inner webs thickly vermiculated with black. The inner and longest scapulars have both webs marked in this manner, their inner webs being the darker. The last of the longer scapulars with an irregular terminal spot of black. Lesser wing coverts nearly black tipped with chocolate brown. Bases of primary coverts fulvous brown, the outer webs heavily spotted with rufous brown, the inner webs less so, and a subterminal bar of black crossing entire outer and half of inner web, all the feathers tipped with prominent spots of creamy white. Secondary coverts like primary coverts but the black bar and white spot confined to outer web.

Primaries fulvous brown when held toward light, changing to smoky brown when held away from light. Outer webs spotted with buffy white throughout their entire length, the spots being much lighter on second and third primaries. Tips of feathers mottled with rufous brown. General color of secondaries same as primaries, their outer webs and tips being spotted with rufous brown and these spots in turn being speckled with fulvous brown. Inner three secondaries speckled with fulvous brown, rufous brown and creamy white, each feather with a terminal spot of fulvous. General color of tail rufous brown distinctly barred with lighter rufous brown, each of these bars succeeded by a narrow irregular bar of black, the entire feather thickly speckled with black and each feather having a small black terminal spot. Throat and fore-breast like sides of face. A buffy white pectoral band continuous with nuchal collar and succeeded by a second creamy white band, the feathers between the two bands being brown thickly vermiculated with black and creamy white. Abdomen lighter. Flanks and under tail coverts ashy, slightly tinged with pinkish, some of the feathers with dark black shaft stripes, others with small terminal spots of black. Under surface of tail much like coverts, the black markings of upper surface showing only faintly. Shafts of tail-feathers creamy white. Under wing coverts fulvous brown tipped with white. Axillaries white. Eyes pale yellow. Legs, feet and nails nearly white. Upper mandible brown, lower dirty green. Culmen, 1.05 inches. Wing, 5.46. Tail, 4.14. Tarsus, .61.

Food, beetles. Native name "cow-cow." The single specimen obtained is a fully adult male. Its rich and complicated markings are very difficult to describe.

We have named it in honor of Mr. Menage.

8. Ceyx nigrirostris *sp. nov.*

Adult male: General color of back and upper tail-coverts bright cobalt blue, slightly lighter than in C. cyanipectus. Crown and nape blue-black thickly spotted with bright cobalt, the spots being much wider and slightly lighter than in C. cyanipectus. Spots much larger on hind neck, causing it to appear nearly uniform cobalt.

Scapulars black, heavily washed with dark verditer blue. Wing-coverts washed with verditer blue, each feather with a bright spot or stripe of cobalt blue. Wing black, the outer webs of secondaries heavily washed with light verditer blue. Tail black, the central pair of feathers washed with verditer blue on both webs, the others on outer webs only. Loral spot reddish buff. A spot of same color on sides of neck. Chin and throat white, washed with buff. Fore-neck, breast and abdomen uniform buff. Flanks, sides of breast and a complete band across the breast dark verditer blue. A half band of same color behind this. Under tail-coverts buff, the larger ones tipped with verditer blue. Under wing coverts like the breast, with a spot of verditer blue at end. Basal portion of inner webs of primaries and secondaries washed with pale buff. *Bill black.* Average measurements from ten males: Culmen, 1.42 inches. Tarsus, .34. Wing, 2.22. Tail, .88. Length of a single male measured in the flesh, 6.50. Female like male, but has only a half band of verditer blue across the breast, this being more imperfect than in C. cyanipectus. Average measurements from three females: Culmen, 1.45 inches. Tarsus, .33. Wing, 2.34. Tail, .95. Length of single female measured in the flesh, 5.63 inches.

A well marked species easily distinguished from C. cyanipectus its nearest ally, by the heavier markings on crown and nape, by its black bill and by the entirely different color of its under surface. Like the former species, it is strictly confined to the banks of fresh water streams and it is usually found in the woods.

Habitat: Panay, Negros, Cebu.

9. Centropus steerii sp. nov.

Sexes alike. Forehead, crown and nape, sides of face, chin, throat and upper breast greenish black. The coarse shafts of the feathers shiny black, the webs with a faint greenish tinge. Hind neck and back, sides of neck, wing-coverts and breast smoky brown with faint greenish tinge. Hind back and rump slaty black, tips of feathers with greenish tinge. Upper tail-coverts and upper surface of tail uniform dull metallic green. Shafts of feathers jet black from base to tip. Upper surface of wings earthy brown with metallic green gloss like the tail, except on the outer four primaries, which have little gloss. Abdomen browner than breast and with less metallic wash. Flanks, thighs and under tail-coverts like rump. Under surface of tail black with faint metallic blue gloss. Under wing coverts and axillaries like breast. Under surface of wing uniform blackish brown.

Seven males measure as follows: Length, 16.70 inches. Wing, 5.90. Tail, 8.49. Culmen, 1.58. Tarsus, 1.65.

A female measures 19.50 inches in length. Wing, 6.21. Tail, 9.16. Culmen, 1.74. Tarsus, 1.69.

Habitat: Mindoro.

The strong, hooked beak of this species is very noticeable. A single specimen was collected by the Steere Expedition, and Dr. Steere has held a manuscript description of it ever since. We obtained a fine series of specimens, and the doctor made over his claim to us. We are indebted to Dr. Steere for the use of much valuable material for comparison and name this species in his honor. It is invariably found in deep forests where it is not uncommon.

10. Iyngipicus menagei sp. nov.

Adult male: General color of upper surface dark blackish brown. Top of head uniform with back. A small spot above and behind eye creamy white. Scarlet stripes on sides of occiput shorter than in I. maculatus and beginning farther back. They are confluent on nape. Behind and under the scarlet stripe is a partially concealed spot of creamy white. Scapulars, interscapulars and back barred with creamy white. Rump creamy white, some of the feathers with narrow brownish black shaft stripes. Upper tail-coverts brownish black, broadly edged with buffy white. Tail brownish black, paler at base of feathers and with both webs of feathers spotted with pale buff. Wing coverts brownish black, each feather having one or two creamy white spots on outer web. Wing brownish black. Outer five primaries with two or three very narrow creamy white spots on outer web, or with no spots at all. Tips of inner primaries and inner webs of all primaries spotted with creamy white. Second-aries similarly spotted on both webs. Ear-coverts rusty brown. A creamy white malar stripe extending back of ear-coverts. Chin and narrow stripe down center of throat white, bordered by a broad stripe of brownish black on each side, the tips of feathers forming side stripes being brownish white. Under surface with strong ful-vescent wash. Feathers of upper breast with distinct brownish black shaft marks. Feathers of lower breast and abdomen with ill defined streaks of the same color. Feathers of flanks nearly white, with only slight dark markings. Under tail-coverts yellowish white, with dark shaft stripes. Under surface of tail slightly lighter than upper, but tips of two central pairs of feathers nearly black. Under wing-coverts and axillaries creamy white, spotted with brownish black. Bend of wing uniform brownish black.

The female lacks the scarlet head markings of the male and the creamy white spot, which is partially concealed in the male, is in the female quite conspicuous. Otherwise the sexes are alike.

Five males measure in length, 5.84 inches. Culmen, .80. Wing, 3.07 Tail, 1.59. Tarsus, .59. Eight females: Length, 5.97. Culmen, .79 Wing, 3 19. Tail, 1.63. Tarsus, 66. Habitat: Sibuyan.

11. Chibia menagei *sp. nov.*

Adult male in worn out plumage. Black. Wings with metallic green gloss. Feathers of head, nape, neck and breast with spangles of metallic blue, broad on crown, elsewhere narrow. Scapulars, interscapulars, back and rump blue-black, with faint metallic gloss. Upper tail coverts more strongly glossed, especially on outer web. Central tail-feathers and outer webs of others glossed like the wings. Tail graduated, the outer pair of feathers exceeding the next inner pair by one and one fourth inches and strongly curved upward and in-ward, so that at tip the inner web of feather is turned outward. Neck-hackles considerably elongated. Female like male. None of the specimens show frontal plumes.

Average measurements of nine males: Length, 13.25 inches. Cul-men, 1.37. Wing, 5.39. Tail, 6.91. Tarsus, 1.

Of seven females: Length, 12.87. Culmen, 1.37. Wing, 5.36. Tail, 6.69. Tarsus, 1.01.

This curious species is by far the largest representative of its genus yet discovered in the Philippine Islands, and differs striking-ly from both the other known species, one of which is confined to Palawan and the Calamianes Islands, while the other occurs in the Sulu group and in Cagayan Sulu.

C. menagei seems to be strictly confined to the island of Tablas where it is not rare in the deep woods.

12. Oriolus cinereogenys sp. nov.

In uniting the Tawi Tawi birds with O. steerii from Basilan and Mindanao, Dr. Sharp has evidently overlooked the fact that the Tawi Tawi birds invariably have the cheeks and ear-coverts clear ashy grey, while in birds from Basilan and Mindanao they are just as invariably olive green. As we find no exception to this rule among our fourteen specimens from Basilan and twenty from Tawi Tawi we have no hesitation in separating the birds from the latter locality. It may be added that the rump of the Tawi Tawi birds is rather brighter, and the throat decidedly lighter than in Basilan birds. Not one of our Tawi Tawi birds shows the uniform grey throat of O. steerii. Both species show great variability in the color of under tail coverts. In some specimens they are pure yellow and in others heavily streaked with black.

Fifteen males from Tawi Tawi measure as follows: Length, 8.01 inches. Culmen, .96. Wing, 4.59. Tail, 3.19. Tarsus, .83.

Habitat: Tawi Tawi.

13. Oriolus nigrostriatus sp. nov.

Similar to O. steerii, from which it differs in having the lores, chin, throat and upper breast decidedly darker ashy and the mesial stripes of feathers of breast and abdomen broader and much deeper black, the general color of wing darker and the washing on inner webs of quills white instead of yellow. Rump yellower than in O. steerii, the edges of feathers of rump bright yellow. Wing more like that of O. assimilis than that of O. steerii, only a few of the secondaries and tertiaries having any wash of yellowish green. Lower primary coverts have no yellow wash. Sexes alike. Length, 8.75 inches. Culmen, 1.08. Wing, 4.67. Tail, 3.60. Tarsus, .82.

The first specimen of this species was obtained by Dr. Steere in the island of Negros in 1874, and was described by Dr. Sharpe Trans. Linn. Soc. (2) Zool. i, p. 329 (1877), who called attention to certain differences between it and O. steerii from Basilan but did not care to found a species on such slender evidence as he had at hand. The Steere Expedition obtained a single specimen in Masbate, which Dr. Steere incorrectly identified as O. assimilis, a mistake which could not have occurred had he had any specimens of O. assimilis for comparison. Strangely enough the species is far more closely allied to O. steerii than to its geographically much nearer neighbor in Cebu.

Habitat: Negros, Masbate. It will doubtless eventually be discovered in Panay also.

14. Aethopyga arolasi sp. nov.

Adult male: Slightly larger than Aethopyga bella. Upper surface as in that species. Fore-breast much more broadly streaked
with orange. Abdomen and under tail-coverts light lemon yellow,
not white.

Adult female: Above uniform olive green. Does not show the
bright yellow rump of Aethopyga bella. Under surface inclin-
ing to white, but breast, abdomen and under tail-coverts washed
with light lemon yellow.

Average measurements from ten males: Length, 3.44 inches.
Culmen, .70. Wing, 1.59. Tail, 1.30. Tarsus, .46.

From four females: Length, 3.26 inches. Culmen, .63. Wing,
1.59. Tail, .99. Tarsus, .49.

Habitat: Tawi Tawi and Sulu.

We have named this beautiful sun-bird in honor of Brigadier General Juan Arolas, for many years governor of the Sulu group, to
whom we are indebted for much personal kindness and for assistance without which our work in Sulu would have been almost impossible.

15. Aethopyga bonita sp. nov.

Above as in Aethopyga arolasi except that the rump is orange
yellow instead of sulphur yellow and the metallic spot on forehead
is violet instead of metallic green when held away from the light.
Chin, throat and fore-breast bright orange yellow, thickly streaked
with deep orange red. Mustachial line, sides of face and ear-patch
as in Aethopyga arolasi. Lower breast, flanks, abdomen and under
tail-coverts white, distinctly washed with lemon yellow. Under
wing-coverts and inner webs of quills white. Under surface of tail
black, tips of some of the feathers grey.

Adult female: Above olive green, nearly brown on forehead and
crown. Rump bright yellow as in Aethopyga bella. Tail and quills
brownish black washed with olive brown. Sides of face and ear-
coverts ashy grey. Sides of neck olive green. Chin and throat
nearly white. Fore-breast much darker. Hind-breast, abdomen,
flanks and under tail coverts whitish, strongly washed with pale
yellow. Under wing-coverts, axillaries and inner webs of quills
white.

Four males measure as follows: Length, 3.72 inches. Culmen,
.72. Wing, 1.72. Tail, 1.28. Tarsus, .56.

A female measured 3.50 in length.   Culmen, .68.   Wing, 1.58.
Tail, 90.   Tarsus, .52.

Habitat: Negros, Cebu, Masbate.

This pretty species can be readily distinguished from the last, its
nearest ally, by its orange yellow throat with heavier orange red
markings. The female also differs from that of Aethopyga arolasi
in having a yellow rump.

16. Aethopyga minuta sp. nov.

Adult male. Like Aethopyga arolasi, but smaller, and the throat
pure yellow without the faintest trace of orange red. We regret
that we are unable to furnish exact measurements because our
type specimen, a male in fine plumage disappeared from the collection in July, 1894.

Habitat: Mindoro.

This tiny species is rare in Mindoro and was seen by us on but
two occasions.

17. Dicaeum pallidior sp. nov.

Adult male. Above exactly like D. dorsale, which by the way has
the back slaty blue, not slaty grey, as stated in Cat. B., Vol. x , p.
40. Entire under surface yellow, only slightly deeper on the breast,
and not rich orange as in D. dorsale. Habitat: Cebu.

Were it not that we have a large series of birds from Cebu in
breeding plumage we should not think of separating the Cebu birds,
but with the series of specimens now before us we cannot do otherwise. The almost uniform yellow under surface of D. pallidior,
contrasts strongly with the yellow throat and abdomen and bright
orange breast of D. dorsale. Young males of D. dorsale show
streaks of orange on the breast long before reaching maturity, hence
the difference is not a matter of age and, as we have already re-
marked, our Cebu birds were in breeding plumage. The female is
like that of D. dorsale.

Measurements of eleven males: Length, 3.50 inches. Culmen, .55.
Wing, 1.92. Tail, 1.03. Tarsus, .51.

Of four females: Length, 3.31. Wing, 1.88. Tail, .98. Tarsus, 49.

18. Dicaeum sibuyanica sp. nov.

A well marked species of the D. dorsale type and the largest representative of this type yet discovered in the Philippine islands. A
very noticeable characteristic is that fully adult birds always have
the base of the lower mandible whitish as do the young of most
other species of the genus.

Adult male: Upper surface as in D. dorsale but the rump quite heavily washed with olive green. Chin, throat and upper breast light bluish ashy grey, slightly paler on the chin and without the yellow of D. besti. Two of our specimens, however, show a faint trace of yellow on the chin. Rest of under surface yellow, slightly deeper on breast as in D. pallidior. Flanks, abdomen and under tail coverts much paler. Under wing-coverts, axillaries and inner webs of quills white.

Adult female: Upper surface as in D. besti. Chin, throat and upper breast grey, washed with yellow. Rest of under surface pale greenish yellow, somewhat brighter along center of breast and abdomen, but not nearly as bright as in female of D. besti. Bill as in male.

Average measurements from nine males: Length, 3.9 inches. Culmen, .59. Wing, 2.08. Tail, 1.03. Tarsus, .53. From two females: Length, 3.81. Culmen, .61. Wing, 1.96. Tail, .97. Tarsus, .55.

Habitat: Sibuyan.

19. Dicaeum intermedia sp. nov.

Adult male: Above as in D. dorsale. Rump shows very little olive green wash. Chin and throat ashy grey uniformly washed with pale yellow. Remainder of under surface as in D. sibuyanica. Bill black.

Adult female: Above like female of D. sibuyanica. Below dirty olive yellow, somewhat brighter on the abdomen. Bill paler than in male.

Habitat: Romblon, Tablas. It may ultimately prove that the Tablas birds are distinct, the four specimens secured by us in that island having a much heavier wash of yellow on the throat than the Romblon birds.

Five males from Romblon measure 3.75 inches in length. Culmen, .55. Wing, 2.08. Tail, 1.09. Tarsus, .52. Three females from Tablas measure 3.62 inches in length. Culmen, .55. Wing, 1.92. Tail, 1.02. Tarsus, .52.

20. Dicaeum assimilis sp. nov.

Adult male: Above exactly like D. sibutense, but chin, throat and fore breast very much lighter than sides of face, being clear ashy grey, as in D. trigonostigma. As Dr. Sharpe expressly states that this is not the case in D. sibutense but that the latter species has the throat like the sides of the face and head. it is evident that the Sulu and Tawi Tawi (?) birds belong to a distinct species having the back of D. sibutense and the under surface of D. trigonostigma.

Female like that of D. sibutense, but with the throat, upper breast and sides of face light ashy grey, uniformly washed with yellow. The plate in Ibis seems to show a faint orange mark on the back of the female of D. sibutense, though no such marking is described in the text. If this is the case the female of D. sibutense differs from that of every other Philippine representative of the genus.

A male from Sulu measures 3.50 inches in length. Culmen, 53. Wing, 2.04. Tail, .97. Tarsus, .53.

Two females from the same locality measure 3.31 in length. Culmen, .50. Wing, 1.98. Tail, .97. Tarsus, .48.

Habitat: Sulu, Tawi Tawi (?). We have only females from Tawi Tawi and cannot be quite sure of the identification.

21. Prionochilus aeruginosus *sp. nov.*

Adult male in poor plumage. Upper surface light rusty brown, washed with olive yellow on rump, upper tail coverts, outer webs of secondaries and tail feathers. Ear coverts and sides of neck lighter than crown. Lores, like side of head, bordered above and below by narrow, creamy white stripes reaching to eye. A creamy white malar stripe is separated from these and from the throat by light brown stripes of about the same width. Throat white. All the feathers with broad brown shaft markings extending to their tips give the breast a strongly striped appearance. Stripes become less sharply defined on the flanks and disappear almost entirely on the abdomen. Under tail-coverts creamy white with dark ill-defined brown spots. Under wing-coverts and axillaries creamy white. Inner webs of quills ashy grey, outer webs and tips blackish brown. Female like the male but with less olive yellow above. Young like female but lacking the distinct markings on the under surface.

Measurements from three males: Length, 4.25 inches. Culmen, .45. Wing. 2.60. Tail, 1.42. Tarsus, .57.

A female measures 4.00 in length. Culmen, .53. Wing, 2.44. Tail, 1.28. Tarsus, .53. Habitat: Cebu, Mindanao.

22. Prionochilus bicolor *sp. nov.*

Adult male: Entire upper surface deep black, with faint metallic blue gloss. Entire under surface, including under wing-coverts, axillaries and inner webs of quills, white. Bases of feathers of breast, flanks and abdomen slaty black. Bill, legs and feet black. Iris red. Length, 3.25 inches. Wing, 1.98. Tail, .96. Culmen, .42. Tarsus, .56. Habitat: Mindanao.

Found in the hills back of Ayala, near Zamboanga.

23. Zosterops siquijorensis *sp. nov.*

Adult male: Above light olive yellow, brighest on crown and rump. Wing-coverts and broad margin on all the quills except last uniform with the back. Tail brown, central pair of feathers washed on both webs, and the others on the outer webs, with olive yellow. Ear-coverts, sides of face and sides of neck slightly lighter than crown of head Ring around eye silky white. No black stripe under eye as in Z. everetti. Lores and forehead bright yellow. Chin, throat and upper breast bright yellow as in Z. meyeni. Breast, flanks and abdomen clear ashy grey, lighter than in Z. everetti. A narrow stripe of yellow down center of breast and abdomen. Under tail-coverts bright yellow. Under wing-coverts, axillaries and inner webs of quills white. Bend of wing tinged with yellow. Sexes alike. Iris brown. Legs and feet very light brown. Upper mandible brown. Lower yellowish.

A male from Siquijor measures as follows: Wing, 2.16. Tail, 1.51. Culmen, 1.55. Tarsus, 1.57.

Two females measure: Length not taken. Wing 2.13. Tail 1.53. Culmen, .56. Tarsus, .56. Habitat: Siquijor.

Singularly enough the Siquijor Zosterops differs sharply from the Zosterops of the neighboring island, Cebu, and more nearly resembles Z. meyeni of Luzon. It differs from Z. meyeni in its larger size and in the different color of the under surface.

24. Hyloterpe winchelli *sp. nov.*

Adult male: General color of upper surface umber brown. Head faintly but appreciably darker. Outer webs of quills washed with light umber. Rest of quills dark fulvous brown. Tail like back, or slightly darker. Shafts of tail-feathers become darker on terminal half. Lores. sides of face, ear-coverts and sides of hind neck slightly lighter than head. Chin and throat greyish white. Sides of breast and fore-breast washed with light umber brown, as are the flanks. Rest of under surface white. Feathers of thighs dark brown, heavily tipped with white. Under surface of tail dark brown. Shafts white for entire length. Under wing coverts and axillaries and inner webs of quills white. Iris brown. Bill black. Legs and feet light slaty brown.

Average measurements of six males: Length, 6.50 inches. Wing, 3.24 Tail, 2.66. Tarsus, .75 Sexes alike.

Habitat: Panay, Masbate, Negros.

This species is named in honor of Mr. Horace V. Winchell, through whom interest in our proposed expedition was first awakened.

25. Hyloterpe major *sp. nov.*

Adult male: General color exactly as in H. winchelli from which it differs in having the white of the throat less sharply defined and in its much larger size.
Four males from Cebu measure 7.31 in length. Wing, 3.70. Tail. 2.97. Culmen, .80. Tarsus, .88.
Three females from the same locality measnre 7.00 inches in in length. Wing, 3.69. Tail, 2.83. Culmen, .79. Tarsus, .87.
Three males from Tablas measure in length 7.00 inches. Wing. 3.40. Tail, 2.79. Culmen, .75. Tarsus, .89.
Habitat: Cebu, Tablas, Sibuyan.
It will be noticed that the Tablas birds are slightly smaller than those from Cebu. The same holds true of the Sibuyan birds. We note also that the Cebu birds have the feathers of the thighs dark brown faintly tipped with white while Sibuyan birds have the same feathers light brown heavily tipped with white. The under tail-coverts in the Cebu birds are usually not white but are washed with brown, in one case heavily; but we note a good deal of variation in the color of under parts in birds from the same locality, also that the largest of our Tablas-Sibuyan birds are larger than the smallest of the Cebu birds, hence do not think they can be specifically separated. The distribution is curious, as H. winchelli comes between; but this is only one of several instances of relationship between Cebu and the Tablas-Romblon-Sibuyan group.

26. Hyloterpe mindorensis *sp. nov.*

Adult female.   Above brown washed with olive yellow, faintly on head, more heavily on back, the rump bright olive yellow.   Scapulars, wing-coverts and outer webs of tertiaries heavily washed with olive yellow changing to olive brown on secondaries and primaries. Tail olive yellow above and below, brighter on basal half.   Shafts of feathers above brown, below bright yellow.   Lores and sides of face ashy brown.   Ear-coverts with distinct light shaft stripes. Chin, throat and upper breast ashy grey, changing to olive brown on sides of neck, breast and upper breast.   Lower breast and abdomen yellowish white.   Flanks grey washed with olive brown. Under tail coverts light yellow.   Axillaries, under wing-coverts and edges of inner webs of quills whitish.   Length, 6.50 inches. Culmen, .70. Wing, 3.09. Tail, 2.60. Tarsus, .84.
Habitat: Mindoro.

27. Cryptolopha flavigularis *sp. nov.*

Adult male. Above bright olive green. Head like back. All the wing feathers broadly edged with olive green. Tail olive green above, the inner webs of the outer three pairs of feathers narrowly edged with bright yellow. Shafts dark brown above. A bright yellow superciliary line extending from base of bill to ear. A similar line below eye and not extending beyond it. Lores olive brown. Sides of face and ear-coverts light olive yellow with lighter yellow shaft markings. Chin, throat and upper breast white heavily washed with light yellow. Rest of under surface whitish streaked with light yellow. Under tail coverts yellow, shafts white. Under surface of tail olive green, shafts pure white. Axillaries, under wing coverts, inner edge of quills, yellowish white. Bend of wing light yellow. Length, 5.00. Wing. 2 29. Tail, 1.88. Culmen, .63. Tarsus, .72.

Legs, feet and nails very light brown. Upper mandible dark brown, lower light brown.

Habitat: Cebu.

The specimen described is in breeding plumage.

28. Geocichla cinerea *sp. nov.*

Entire upper surface very dark ashy, nearly black. Tail and tips of wing feathers rusty brown. Wing-coverts brownish black, each tipped with a large spot of white, these spots forming two irregular wing bars. Lores whitish. Spot under eye brownish black. Chin and center of throat white. A broad line of brownish black extends from the gape to the breast where it joins a large patch of black formed by feathers the bases of which are white. Rest of under surface white, the feathers of sides of breast and flanks with narrow black tips. Center of lower breast and abdomen as well as under tail coverts unspotted. A slight tawny wash on flanks. Axillaries white tipped with ashy grey. Under wing coverts ashy grey tipped with white. A prominent spot of white on inner web of each secondary, the spots together forming a patch. Bend of wing white. Wing, 4.40. Tail, 2.86. Culmen, .91. Tarsus, .98.

Habitat: Mindoro.

29. Cittocincla superciliaris *sp. nov.*

Adult male: Entire upper surface, sides of head, chin, throat and fore-breast glossy black. A broad superciliary stripe of pure white begins in front of eye, tapering to a point on the hind-neck where it approaches, but does not meet, the stripe of other side.

Central pair of tail-feathers shows faint traces of bars by reflected light. Under surface pure white. Bend of wing white. Outer tail-feathers, which are but half grown, tipped with white. Iris very dark brown. Legs and feet almost white. Bill coal black.

A young male, nearly adult, has a few white feathers on chin and throat and a faint wash of light buff on the flanks. A much younger bird has many of the feathers of the back tipped with rusty brown and the greater wing coverts and quills washed with the same color. Chin and throat almost pure white. An ill defined black collar. The entire under surface washed with light buff, deeper on the flanks.

Adult male: Length, 6.87. Wing, 3.10. Tail, 2 80. Culmen, .76. Tarsus, .86.

Habitat: Masbate.

This well marked species is extremely rare in Masbate. It feeds in dense thickets in the deep woods and we never heard it utter a note. The Luzon bird, C. luzoniensis, has a superciliary stripe but this stripe is not nearly so broad as in this species and as the strongly marked superciliary line is one of the most noticeable characters of the Masbate bird we have named it accordingly.

30. Ptilocichla minuta *sp. nov.*

Sexes alike. Feathers of the head and nape black, with heavy rufous brown shaft lines. Feathers of back and upper wing coverts bright reddish brown, with conspicuous nearly white shaft markings for their entire length. Tips of feathers black. The elongated feathers of back which reach to tail-coverts with white shafts and white shaft markings broad at base and narrowing at tip, edges and extreme tips of feathers being dark rich fulvous brown. Upper tail coverts rufous brown. Tail feathers fulvous brown edged with rufous brown. Wing feathers rufous brown. Lores white. A superciliary line of white extending as far as hind neck. Ear-coverts fulvous with light shaft stripes, the latter becoming rufous on hind-neck. A malar stripe of black. Chin and throat pure white. Feathers of breast and abdomen have very broad white shaft stripes, giving a streaked appearance to the under surface. Feathers of flanks much elongated, light fulvous brown with distinct white shaft stripes, broadest at base. Under tail-coverts colored like flanks. Under surface of wing fulvous brown, brighter on coverts.

Readily distinguished from P. basilanica by having all the feathers of back, head, rump and upper wing-coverts with prominent shaft lines, by the darker color of the long feathers of the back and its much smaller size. Sexes alike. Exceedingly rare. Length, 5.37. Wing, 2.71. Tail, 1.64. Culmen, 7.41. Tarsus, .96 Habitat: Samar.

31: Iole cinereiceps *sp. nov.*

Sexes alike. Pointed feathers of head olive brown at base, strongly washed with ashy grey, and with nearly black shaft stripes, giving the whole upper surface of the head a nearly uniform grey color. Sides of face, ear coverts, hind neck, back, rump and upper wing-coverts olive brown. Upper t·il-coverts and tail blackish brown, edges of feathers washed with olive brown. Chin and throat yel-lowish white, the feathers with very distinct pure white shaft mark-ings. Fore-breast and sides of breast light olive brown, the feathers with distinct white shaft markings. Flanks light olive brown with faint white shaft markings. Center of breast, abdomen, and under tail-coverts uniform yellowish white. Under surface of tail ashy white, shafts pure white. Under wing coverts and axillaries yel-lowish white, darker than abdomen. Under surface of quills like under surface of tail.

Average of nineteen males: Length. 11.33 inches. Wing, 5.15. Tail, 4.53. Tarsus, .88. Culmen, 1 39. Of four females: Length, 10.78. Wing, 4.80. Tail, 4.29, Culmen, 1.39. Tarsus, .86.

Iris dark brown. Legs, feet and nails dark brown. Bill dark brown in some specimens, black in others.

Habitat: Tablas, Romblon.

This fine Iole is the largest Philippine representative of the genus. Its nearest allies are Iole siquijorensis and Iole monticola. It is a woods bird very rarely met with in the open.

32. Iole monticola *sp. nov.*

Adult male. Crown and nape blackish brown, feathers of fore-crown and forehead washed with ashy grey. Upper surface of body rusty brown. Feathers of back and scapulars with distinct lighter shaft markings. Outer quills of wing washed with rusty brown. Rest of wing light fulvous brown. Feathers of lower back, rump and upper tail-coverts distinctly washed with olive green. Feathers of tail like wing but faintly edged with olive green on outer webs. Shafts light brown. Lores whitish. Ear coverts rufous brown with distinct lighter shaft markings. Sides of face somewhat lighter but also with distinct shaft markings. Chin and throat tawny white, the feathers with light shaft markings. Fore-breast and sides of breast rusty brown fading into tawny white on breast, all the feathers with well defined white shaft markings. Abdomen creamy white. Flanks and under tail coverts light fulvous brown. Under surface of tail greyish white, shafts pure white. Axillaries and under wing-coverts yellowish white. Under surface of wing like that of tail. Darker at tip of feathers. Iris, legs, feet and nails dark brown. Bill nearly black. Four males average 8.62 inches in length. Wing, 3.94. Tail, 3.53. Culmen, .98. Tarsus, .72. Four females measure 8.09 inches in length. Wing, 3.69. Tail, 3.29. Culmen, .98. Tarsus, .70. Habitat: Cebu.

Iole monticola differs from Iole siquijorensis its nearest ally in its more ruddy upper surface, its lighter head with a wash of ashy grey on front of crown, in its lighter ear-coverts and tawny throat and in its lighter under wing and tail-coverts.

So far as our observation goes it is a highland form. It was invariably met with by us in the forest on the tops and sides of the hills in central Cebu and was never seen in open or flat country.

33. Muscicapula samarensis *sp. nov.*

Adult male: Top of head, sides of face, ear-coverts and hind neck nearly black. Back, rump and upper wing-coverts uniform dark slaty blue. Quills and tail fulvous brown slightly washed with slaty blue. Chin and throat white. Entire breast bluish grey. lightest on center of breast. Abdomen white. Flanks washed with bluish grey. Under wing-coverts light buff nearly white at base. Sides dark slate color as are under wing-coverts and axillaries the latter, however, mottled with white. A superciliary stripe of white beginning over eye and extending to nape, then inward, nearly reaches the median line. Sexes alike. Iris very dark brown. Bill black. Legs, feet and nails very light brown. Measurements from four males: Length, 4.67. Wing, 2.41. Tail, 1.49. Culmen, .59. Tarsus, .79.

The specimens described are in breeding plumage. They were shot close to, or on the ground in dense thickets in the deep woods.

This species is closely allied to M. mindanensis Blas from which it differs in its darker head, lighter tail, and much larger superciliary stripe. None of our specimens show a white bar on the rump but we find the Mindanao Basilan birds variable in this respect.

Habitat: Samar.

34. Rhipidura sauli *sp. nov.*

Adult male: Head, crown and nape dull bluish grey, each of the feathers of crown with a narrow decidedly lighter shaft mark, lacking in feathers of nape and mantle. Back, rump, upper tail-coverts, scapulars and upper wing-coverts chestnut. Wing black. Tertiaries and secondaries heavily edged with chestnut. Primaries lightly edged with same color. Central pair of tail-feathers black, faintly edged with chestnut on basal half and with shafts black. Next pair have inner webs black, outer webs chestnut, shafts black on inner side, chestnut on outer. Rest of feathers of tail including shafts chestnut above and below. Sides of face, ear-coverts, chin, throat and upper breast bluish grey like the mantle. Feathers of breast with distinct lighter shaft-markings. Feathers of lower breast gradually changing to the chestnut of abdomen. Flanks, under tail-coverts and thighs chestnut. Axillaries and under wing-coverts bluish grey at base, heavily tipped with chestnut. Inner webs of quills tipped with chestnut. Female like male but paler. Iris dark brown. Legs and feet vary from light to very dark slaty brown. Nails blackish. Bill black, except base of lower mandible which is grey. Twelve males average 7.25 inches in length. Wing, 3.30. Tail, 3.59. Culmen, .65. Tarsus, .71. Three females: Length, 6.87. Wing, 3.00. Tail, 3.27. Culmen, .64. Tarsus, .74.

Habitat: Tablas.

Another ornithological puzzle of the Tablas-Romblon-Sibuyan group. It seems to be confined to Tablas where it is common in deep forests. It differs from R. cyaniceps, its nearest ally, in its larger size and darker blue head and in having the ochraceous buff of under parts replaced by deep chestnut. We have named this species in honor of our friend Geo. M. Saul Esq., of Ilo Ilo, to whom we are greatly indebted for much kindness and courtesy shown to us during both of our visits to the Philippine islands.

35. Rhinomyias albigularis *sp. nov.*

Adult male: General color above ochraceous brown, duller on head, much brighter on rump, becoming chestnut on the tips of upper tail-coverts. Upper wing-coverts like back. Quills nearly black washed with rusty brown on outer webs, this wash changing to whitish on the primaries. Upper surface of tail dull chestnut, the feathers becoming almost black at tips. Lores grey. Ear-coverts and sides of hind-neck like crown. A ring of feathers around eye slightly lighter. Chin and entire throat white. Entire breast light olive brown. Flanks washed with same color. Abdomen pure white. Under tail-coverts white, light brown at tips. Under wing-coverts, axillaries and inner webs of quills buffy white. Bend of wing olive brown. Length in skin — — inches. Culmen, .71. Wing, 3.47. Tail, 2.68. Tarsus, .75.

Habitat: Negros, Guimaras.

The white throat contrasts strongly with the brown of neck and breast and at once distinguishes this species from all other Philippine representatives of the genus. R. albigularis is a deep woods form and is extremely rare in the localities visited by us.

36. Rhynomyias ocularis *sp. nov.*

Adult male: General color above uniform olive brown, slightly brighter on the rump. Tail dark chestnut, much brighter on outer webs of feathers which are very dull at tips.   Wing-coverts like back.   Quills brownish black washed with rusty brown, this becoming whitish on first two or three primaries.   Lores buffy white. A ring of short feathers around eye chestnut. Ear-coverts and sides of hind-neck fulvous brown, the former with lighter shaft stripes.   Center of throat and fore-breast white, greyish along sides. Breast and flanks washed with light fulvous brown.   Abdomen and under tail-coverts white, the latter faintly tipped with brown. Thighs olive brown.   Under wing-coverts and axillaries whitish. Inner webs of secondaries edged with buffy white. Sexes alike. The peculiar ring of feathers around the eye forms a noticeable character by which this species is readily distinguished from other Philippine representatives of the genus.   Iris brown.   Bill light slaty brown. Feet slaty brown, nails darker.   Sexes alike.   Measurements from five males: Length, 5 97 inches.   Wing, 2.97.   Tail, 2.60.   Culmen, .69.   Tarsus, .72.   From four females: Length, 6 08. Wing, 3.09.   Tail, 2.60.   Culmen, .72.   Tarsus, .73.

Habitat: Sulu, Tawi Tawi.

Food usually insects but two specimens had been eating fruit when shot.

Feeling an especial interest in the Zoo-geographical problems suggested by the previously ascertained facts in regard to the distribution within the group of the resident land birds we worked out as carefully as possible the distribution of all the species of birds and mammals met with.   Following is a list of the species for which new localities were determined.   As has already been stated, Mr. A. Everett visited Tawi Tawi some months after our departure from that island.   Species found there by him as well as ourselves are marked with a star placed after the name of the island.   We found Dr. Platen in Mindoro on our arrival there and it would seem that the large collections made by him in this interesting island must long since have been described, but we have been unable to ascertain where and must therefore apologize in advance if we seem to claim credit for any of his discoveries. Any corrections as regards the matter of priority will be gladly received and incorporated, with due acknowledgment, in our later and more complete paper.

III.

NEW LOCALITIES FOR SPECIES PREVIOUSLY KNOWN FROM THE PHILIPPINE ISLANDS.

1. Megapodius cumingi Dillwyn; Grant, Cat. B. Brit. Mus. xxii. p. 449. (1893). Tawi Tawi, Sulu, Samar. Tablas, Romblon, Sibuyan.

2. Excalfactoria lineata (Scop.); Grant, Cat. B. Brit. Mus. xxii, p. 253 (1893). Panay, Cebu, Masbate, Calamianes.

3. Gallus gallus (Linn.); Grant, Cat. B, Brit. Mus. xxii, p. 344. (1893). Tawi Tawi, Calamianes, Tablas, Romblon, Sibuyan, Masbate, Negros.

4. Turnix fasciata (Temm.); Grant, Cat. B. Brit. Mus. xxii, p. 535 (1893). Masbate, Panay, Sibuyan, Calamianes.

5. Osmotreron vernans (Linn.); Salvad., Cat. B, Brit. Mus xxi, p. 60. (1893). Negros, Masbate, Calamianes.

6. Osmotreron axillaris (G. R Gr.); Salvad., Cat. B. Brit. Mus., xxi, p. 48, pl. iv. (1893). Tawi Tawi, Siquijor.

7. Phabotreron nigrorum Sharpe; Salvad., Cat. B. Brit. Mus., xxi, p. 68. (1893). Masbate, Tablas, Sibuyan.

8. Phabotreron brevirostris Tweed.; Salvad., Cat. B. Brit. Mus., xxi, p. 89. (1893). Siquijor.

9. Ptilopus occipitalis (G. R. Gr.); Salvad., Cat. B. Brit. Mus., xvi, p. 72. (1893). Samar, Mindoro.

10. Ptilopus leclancheri (Bp.); Salvad., Cat. B. Brit. Mus., xxi, p. 80. (1893). Tablas, Calamianes.

11. Ptilopus bangueyensis Meyer; Salvad., Cat. B. Brit. Mus. xxi. p. 143 (1893). Tawi Tawi.*

12. Carpophaga aenea (Linn.); Salvadori, Cat B. Brit. Mus. xxi. p. 190 (1893). Samar, Tawi Tawi. Calamianes, Tablas, Sibuyan, Panay, Siquijor.

13. Carpophaga poliocephala G. R. Gr.; Salvad., Cat. B. vol xxi. p. 208 (1893). Panay. Cebu, Samar.

14. Myristicivora bicolor (Scop.); Cat. B. Brit. Mus. xxi. p. 227 (1893). Mindoro, Masbate, Tawi Tawi.

15. Macropygia tenuirostris G. R. Gr.; Salvad., Cat. B. Brit. Mus. p. 346 (1893). Masbate, Mindoro, Tawi Tawi.

16. Turtur dussumieri (Temm.); Salvad., Cat. B. Brit. Mus. xxi. p. 423 (1893). Basilan, Tawi Tawi, Calamianes, Tablas, Romblon, Sibuyan.

17. Chalcophaps indica (Linn.) Salvad., Cat. B. Brit. Mus. xxi. p. 514 (1893). Samar, Tawi Tawi, Calamianes, Mindoro, Siquijor, Ta· blas, Romblon, Sibuyan.

18. Phlogoenas crinigera (Rchnb.); Salvad., Cat. B. Brit. Mus. xxi. p. 587 (1893). Samar.

19. Caloenas nicobarica (Linn.); Salvad., Cat. B. Brit. Mus. xxi. p. 615 (1893). Tawi Tawi, Sulu, Negros. 227 (1875). Siquijor.

20. Hypotaenidia striata (Linn.); Sharpe, Cat. B. Brit. Mus. xxiii. p. 33 (1894). Siquijor, Calamianes.

21. Hypotaenidia torquata (Linn.); Sharpe, Cat. B. Brit. Mus. xiii. p. 43 (1894). Mindoro, Romblon. Siquijor, Masbate.

22. Rallina euryzonoides (Lafresn.); Sharpe, Cat. B. Brit. Mus. xxiii. p. 78, pl. viii. fig. 1. Luzon, Panay, Mindoro.

23. Poliolimnas cinereus (Vieill.); Sharpe, Cat. B. Brit. Mus. xxiii. p. 130 (1894). Luzon, Basilan, Mindoro, Panay, Negros.

24. Amaurornis olivacea (Meyen); Sharpe, Cat. B. Brit. Mus. xxiii p. 153 (1894). Panay, Samar, Cebu.

25. Amaurornis phoenicura (Forst.); Sharpe, Cat. B. Brit. Mus. xxiii. p. 156 (1894). Tawi Tawi, Basilan, Siquijor, Mindoro, Cala· mianes.

26. Gallinula chloropus (Linn.); Sharpe, Cat. B. Brit. Mus. xxiii. p. 169 (1894). Guimaras, Panay, Mindoro.

27. Gallicrex cinerea (Gm.); Sharpe, Cat. B. Brit. Mus. xxiii. p. 183 (1894). Mindoro, Panay, Masbate, Cebu, Mindanao.

28. Porphyrio pulverulentus (Demm.); Sharpe, Cat. B. Brit. Mus. xxiii. p. 507 (1894). Mindoro.

29. Hydrochelidon hybrida (Pall.) Tweed., P. Z. S. 1877, p. 551 Luzon, Samar, Mindanao. Sulu, Tawi Tawi, Calamianes, Tablas, Romblon, Sibuyan, Panay, Guimaras, Negros, Masbate, Cebu. Siquijor.

30. Sterna bergii Licht.; Tweed., P. Z. S. 1877, p. 551. Luzon, Samar, Sulu, Tawi Tawi, Calamianes, Tablas, Romblon, Sibuyan, Panay, Guimaras, Negros, Masbate, Cebu, Siquijor.

31. Sterna sinensis Gm.; Whiteh., Ibis, 1890, p. 60. Mindoro.

32. Sterna leucoptera Meisn. & Schweiz.; Vog., Schweiz, p. 264 (1815). Mindanao.

33. Charadrius fulvus Gm.; Walden, Trans. Zool. Soc. ix. p. 226 (1875). Calamianes, Sibuyan, Masbate.

34. Aegialitis geoffroyi (Wagl.); Wald., Trans. Zool. Soc. ix. p. 227 (1875).

35. Aegialitis dubia (Scop.); Walden, Trans. Zool. Soc. ix. p. 227 (1875). Negros.

36. Aegialitis mongolica (Pall.); Walden, Trans. Zool. Soc. ix. p. 227 (1875). Negros.

37. Strepsilas interpres (Linn.); Tweed., P. Z S. 1878, p. 711. Siquijor, Masbate.

38. Gallinago megala Swinh ; Walden, Trans. Zool. Soc. ix. p. 235 (1875). Tawi Tawi, Calamianes, Sibuyan, Panay, Masbate, Negros, Siquijor.

39. Rhynchaea capensis (Linn.); Walden, Trans. Zool. Soc. ix. p. 235 (1875). Panay, Siquijor.

40. Tringa crassirostris T. & S. Negros.

41. Tringa ruficollis Pallas. Whiteh., Ibis, 1890, p. 59. Negros, Cebu.

42. Tringa subarquata Güldenst. Negros.

43. Tringoides hypoleucus (Linn.); Tweed., P. Z. S. 1877, p 703. Samar, Tawi Tawi, Calamianes, Masbate.

44. Totanus calidris (Linn.); Walden, Trans. Zool. Soc., ix, p. 234 (1875). Cebu.

45. Totanus glareola (Linn.); Blasius, Ornis, 1888, p. 320. Calamianes, Siquijor, Cebu.

46. Terekia cinerea (Güld.); Tweed., P. Z. S. 1878, p. 711. Masbate.

47. Ardea purpurea (Linn.); Wald , Trans. Zool. Soc., ix. p. 338 (1875). Calamianes, Tawi Tawi, Tablas, Mindoro, Palawan.

48. Herodias garzetta (Linn.); Walden, Trans. Zool. Soc., ix. p. 237 (1875). Panay, Siquijor.

49. Herodias intermedia (Wagl.); Walden, Trans. Zool. Soc. ix, p. 237 (1875). Mindoro.

50. Demiegretta sacra (Gm); Tweed., P. Z. S. 1877, p. 551, Sharpe, Ibis, 1894, p. 244. Siquijor, Masbate, Panay, Mindoro.

51. Bubulcus coromandus (Bodd); Walden, Trans. Zool. Soc. ix, p. 237 (1875). Tablas, Masbate.

52. Butorides javanica (Horsf.); Walden, Trans. Zool. Soc. ix, p. 237 (1875). Tawi Tawi, Calamianes, Mindoro, Tablas, Sibuyan, Masbate, Siquijor.

53. Ardetta flavicollis (Lath.); Walden, Trans. Zool. Soc, ix, p. 236 (1875). Cebu.

54. Ardetta cinnamomea (Gm.); Walden, Trans. Zool. Soc. ix, p. 237 (1875). Cebu, Tawi Tawi, Tablas.

55. Ardetta sinensis (Gm.); Walden, Trans, Zool. Soc. ix. p. 237 (1875). Mindoro, Panay, Tablas.

56. Gorsachius melanolophus (Raffl.); Walden Trans. Zool. Soc. ix. p. 238 (1875). Masbate, Cebu, Siquijor.

57. Nycticorax manillensis Vig.; Walden, Trans. Zool. Soc. ix. p. 238 (1875). Samar, Tawi Tawi, Tablas, Sibuyan, Panay, Masbate, Siquijor.

58. Melanopelargus episcopus (Bodd.); Tweed., P. Z. S. 1878, p. 953. Masbate, Panay, Mindoro.

59. Anas luzonica Fraser; Walden, Trans. Zool. Soc. ix. p. 242 (1875). Panay, Masbate.

60. Dendrocygna vagans Eyton; Walden, Trans. Zool. Soc. ix. p. 242 (1875). Mindoro, Panay, Masbate.

61. Fregata minor (Gm.); Whiteh., Ibis, 1890, p. 61. Negros Mindanao.

62. Circus philippinensis Steere; List B. & M. Steere Exped. p. 7 (1890). Negros.

63. Astur trivirgatus (Temm.); Sharpe, Cat. B. Brit. Mus. i. p. 105 (1874). Samar.

64. Accipiter virgatus (Temm.); Sharpe, Cat. B. Brit. Mus. 1, p. 150 (1874). Mindanao, Siquijor.

65. Accipiter manillensis Meyen; Wardlaw Ramsay, Ibis, 1894, p. 330. Guimaras.

66. Spizaetus limnaetus (Horsf.); Sharpe, Cat. B. Brit. Mus. i. p. 272 (1874). Calamianes.

67. Spizaetus philippensis Gurney; Sharpe, Cat. B. Brit. Mus. i. p. 261 (1874). Siquijor.

68. Spilornis holospilus (Vig.); Sharpe, Cat. B. Brit. Mus. i. p. 293 (1874). Sulu, Tawi Tawi. Masbate, Tablas, Romblon, Sibuyan.

69. Spilornis bacha (Daud.); Whiteh., Ibis, 1890, p. 42. Calamianes.

70. Butastur indicus (Gm.); Sharpe, Cat. B. Brit. Mus. vol. i. p. 297 (1874). Calamianes, Tawi Tawi, Sulu, Masbate, Samar.

71. Haliaetus leucogaster (Gm.); Sharpe, Cat. B. Brit. Mus. i. p. 307 (1874). Tawi Tawi, Sulu, Basilan, Luzon, Mindoro, Calamianes, Masbate, Tablas, Romblon, Sibuyan, Siquijor.

72. Haliastur intermedius Gurney; Sharpe, Cat. B. Brit. Mus. i. p. 314 (1874). Basilan, Tawi Tawi, Calamianes, Mindoro, Tablas, Romblon, Sibuyan, Masbate.

73. Elanus hypoleucus Gould; Sharpe, Cat. B. Brit. Mus. i. p. 338 (1874). Tawi Tawi, Calamianes.

74. Pernis ptilonorhynchus (Temm.); Sharpe, Cat. B. Brit. Mus. i. p. 347 (1874). Cebu, Sibuyan.

75. Baza leucopais Sharpe; Whiteh., Ibis, 1890, p. 43, pl. ii. Samar, Romblon.

76. Microhierax erythrogenys (Vig.); Sharpe, Cat. B. Brit. Mus. i. p. 469 (1874). Samar, Mindoro.

77. Falco severus Horsf.; Sharpe, Cat. B. Brit. Mus. i. p. 397 (1874). Cebu, Siquijor, Tawi Tawi, Calamianes, Romblon, Sibuyan.

78. Polioaetus icthyaetus (Horsf.); Sharpe, Cat. B. Brit. Mus. i. p. 452 (1874). Mindoro, Basilan, Calamianes.

79. Ninox lugubris (Tick.); Sharpe, Cat. B. Brit. Mus. ii. p. 154 (1875). Masbate, Sulu.

80. Ninox japonica (P. & S.); Sharpe, P. Z. S. 1879, p. 325. Cebu.

81. Ninox philippensis Bp.; Sharpe. Cat. B. Brit. Mus. ii. p. 167 (1875). Masbate.

82. Strix candida Tick.; Sharpe, Cat. B. Brit. Mus. ii. p. 308 (1875). Calamianes.

83. Eurystomus orientalis (Linn.); Sharpe, Cat. B. Brit. Mus. xvii. p. 33, pl. ii. fig. i. (1892). Tawi Tawi, Calaimanes, Tablas, Romblon, Sibuyan.

84. Pelargopsis gouldi Sharpe; Cat. B. Brit. Mus. xvii. p. 100 (1892). Calamianes.

85. Pelargopsis gigantea Walden; Sharpe, Cat. B. Brit. Mus·
xvii. p. 100 (1892). Tawi Tawi, Cebu, Negros, Tablas, Sibuyan.

86. Alcedo ispida Linn.; Sharpe, Cat. B. Brit. Mus. xvii. p. 141
(1892). Tawi Tawi, Cebu, Negros, Tablas, Romblon, Sibuyan,
Panay.

87. Alcedo meninting Horsf.; Sharpe, Cat. B. Brit. Mus. xvii. p.
157 (1892). Calamianes, Tawi Tawi.*

88. Ceyx euerythra Sharpe; Sharpe, Cat. B. Brit. Mus. xvii, p.
179 (1892). Calamianes, Mindoro, Tawi Tawi.*

89. Ceyx bournsi Steere; Sharpe, Cat. B. Brit. Mus. xvii p. 184
(1892). Tawi Tawi*, Siquijor, Cebu, Negros, Tablas, Romblon,
Sibuyan.

90. Ceyx argentata Tweed.; Sharpe, Cat. B. Brit. Mus. xvii. p.
187 (1892). Basilan.

91. Halcyon coromandus (Lath.); Sharpe, Cat. B. Brit. Mus. xvii. ·
p. 217 (1892). Tawi Tawi, Masbate, Sibuyan.

92. Halcyon gularis (Kuhl.); Sharpe, Cat. B. Brit. Mus. xvii. p.
227 (1892). Basilan, Siquijor, Tablas.

93. Halcyon winchelli Sharpe; Cat. B. Brit. Mus. xxii. p. 255
(1892). Sulu, Tawi Tawi*, Siquijor, Cebu, Tablas, Romblon, Sibuyan.

94. Halcyon pileatus (Bodd.); Sharpe Cat. B. Brit. Mus. xvii. p.
229 (1892). Tawi Tawi.

95. Halcyon chloris (Bodd.); Sharpe, Cat. B. Brit. Mus. xvii. p.
273, pl. vii, fig. iii. (1892). Tawi Tawi, Calamianes, Tablas, Sibuyan,
Masbate, Siquijor.

96. Anthracoceros montani (Oust.); Grant, Cat. B. Brit. Mus.
xvii. p. 370 (1892). Tawi Tawi.

97. Gymnolaemus marchei (Oust.); Grant, Cat. B. Brit. Mus
xvii. p. 270 (1892). Calamianes.

98. Merops bicolor Bodd.; Sharpe. Cat. B. Brit. Mus. xvii. p. 80.
(1882) Samar, Panay, Masbate, Tablas, Calamianes.

99. Merops philippinus (Linn.); Sharpe, Cat. B. Brit. Mus. xvii.
p. 71 (1892). Samar, Tawi Tawi, Sulu, Masbate.

100. Caprimulgus macrurus Horsf.; Hartert, Cat. B. Brit. Mus.
xvi. p. 537 (1892). Sibuyan.

101. Caprimulgus griseatus Gray; Hartert, Cat. B. Brit. Mus.
xvi. p. 550, pl. xi. (1892). Sibuyan.

102.   Lyncornis macrotis (Vig.); Hartert, Cat. B. Brit. Mus. xvi.
p. 605 (1892).   Mindoro.

103.   Lyncornis mindanensis Tweed ; Hartert, Cat. B. Brit. Mus.
xvi. p. 605, pl. xiii.   (1892).   Basilan.

104.   Chaetura gigantea (Temm.); Hartert, Cat. B. Brit. Mus. xvi.
p. 475 (1892).   Mindoro.

105.   Collocalia fuciphaga (Thunb.); Hartert, Cat. B. Brit. Mus.
xvi. p. 498 (1892).   Luzon.

106.   Collocalia francica (Gm.); Hartert, Cat. B. Brit. Mus. xvi.
p. 503 (1892).   Calamianes, Panay.

107.   Collocalia troglodytes Gray; Hartert, Cat. B. Brit. Mus.
xvi. p. 807 (1892).   Samar, Mindoro, Cebu, Siquijor, Masbate, Sibuyan. Romblon.

108.   Collocalia marginata Salvad.; Hartert, Cat. B Brit. Mus.
xvi. p. 508 (1892).   Masbate.

109.   Macropteryx comata (Temm.); Hartert, Cat. B. Brit. Mus.
xvi. p. 517 (1892).   Samar, Sulu, Tawi Tawi, Tablas, Masoate.

110.   Coccystes coromandus. (Linn.); Shelley, Cat. B. Brit. Mus.
xix. p. 214 (1891).   Siquijor, Palawan.

111.   Surniculus velutinus Sharpe; Shelley, Cat. B. Brit. Mus.
xix. p. 230 (1891).   Tawi Tawi*, Sulu. Samar.

112.   Hierococcyx spaverioides (Vig.); Shelley, Cat. B. Brit. Mus.
xix. p. 236 (1891).   Calamianes.

113.   Hierococcyx fugax (Horsf.); Shelley, Cat. B. Brit. Mus. xix.
p. 236 (1891).   Cebu, Basilan, Sulu.

114.   Cuculus canorus (Linn.); Hartert, Cat. B. Brit. Mus. xix. p.
245 (1891).   Siquijor.

115.   Cacomantis merulinus (Scop.); Shelley, Cat. B. Brit. Mus.
xix. p. 268 (1891).   Tawi Tawi*, Calamianes, Tablas.

116.   Chalcococcyx xanthorhynchus (Horsf.); Shelley, Cat. B.
Brit. Mus. xix. p. 289 (1891).   Cebu.

117.   Eudynamis mindanensis (Linn.); Shelley, Cat. B. Brit. Mus.
xix. p. 231 (1891).   Tawi Tawi, Tablas, Romblon, Sibuyan, Panay,
Cebu, Siquijor, Masbate.

118.   Centropus viridis (Scop.); Shelley, Cat. B. Brit. Mus. xix.
p. 349 (1891).   Siquijor, Tablas, Romblon, Sibuyan.

119.   Centropus javanicus (Dumont); Shelley, Cat. B. Brit. Mus.
xix. p. 354 (1891).   Tawi Tawi, Mindanao, Siquijor, Cebu.

120. Dryococcyx harringtoni Sharpe; Shelley, Cat. B. Brit. Mus.
xix. p. 400 (1891). Calamianes.

121. Cacatua haematuropygia (P. L. S. Müll.); Salvad , Cat. B.
Brit. Mus. xx. p. 130 (1891). Calamianes, Panay, Tablas, Siquijor,
Tawi Tawi.

122. Prioniturus discurus (Vieill.); Salvad., Cat. B. Brit. Mus.
xx. p. 417 (1891). Tablas, Sibuyan.

123. Prioniturus cyaniceps Sharpe; Salvad., Cat. B. Brit. Mus.
xx. p. 419 (1891). Calamiaues.

124. Prioniturus verticalis Sharpe; Ibis, 1894, p. 248, pl. vi. figs.
1 and 2. Tawi Tawi*.

125. Tanygnathus luconensis (Linn.); Salvad., Cat. B. Brit. Mus.
xx. p. 424 (1891). Tawi Tawi, Tablas, Romblon, Sibuyan, Siquijor.

126. Tanygnathus burbidgei Sharpe; Salvad., Cat. B. Brit. Mus.
xx. p. 505, pl. xiii. (1891). Tawi Tawi*.

127. Bolbopsittacus intermedius Salvad.; Salvadori, Cat. B. Brit.
Mus. xx. p. 505 pl. xiii. (1891), Samar.

128. Loriculus bonapartei Souancè; Salvad., Cat. B. Brit. Mus.
xx. pp. 530, 619 (1891). Tawi Tawi*.

129. Loriculus regulus Souancé; Salvad., Cat. B. Brit. Mus. xx.
p. 523 (1891). Tablas Romblon, Sibuyan.

130. Xantholaema haematocephala (P. L. S. Müll.); Shelley, Cat.
B. Brit. Mus. xix. p. 89 (1891). Calamianes(?).

131. Xantholaema intermedia Shelley; Shelley, Cat. B. Brit.
Mus. xix. p. 97 (1891). Tablas, Masbate(?).

132. Iyngipicus maculatus(Scop.); Yungipicus maculatus, Steere,
List B. & M. Steere Exped. p. 9 (1890). Cebu.

133. Iyngipicus ramsayi Harg ; Harg., Cat. B. Brit. Mus. xviii.
p 331 (1890). Tawi Tawi.

134. Tiga everetti Tweed.; Harg., Cat. B. Brit. Mus. xviii. p.
418 (1890). Calamianes.

135. Chrysocolaptes erythrocephalus Sharpe.; Harg. Cat. B.
Brit. Mus. xviii. p. 452 (1890). Calamianes.

136. Thriponax javensis (Horsf.); Harg., Cat. B. Brit. Mus.
xviii. p. 498 (1890). Tawi Tawi*, Cebu.

137. Thriponax philippensis Steere; Steere, Ibis, 1891, p. 305.
Panay.

138. Corvus philippinus Bp.; Hartert, J. f. O. 1891, p. 204. Tawi Tawi, Tablas, Romblon, Sibuyan.

139. Sturnia violacea (Bodd.); Sharpe, Cat. B. Brit. Mus. xiii. p. 70 (1890). Mindoro, Tawi Tawi.

140. Sarcops calvus (Linn.); Sharpe, Cat. B. Brit. Mus. xiii. p. 104 (1890). Tawi Tawi, Tablas, Romblon, Sibuyan.

141. Mainatus palawanensis Sharpe; Sharpe, Cat. B. Brit. Mus. xiii. p. 104 (1890). Calamianes.

142. Calornis panayensis (Scop.); Sharpe, Cat. B. Brit. Mus. xiii. p. 147 (1890). Samar, Tawi Tawi, Calamianes, Tablas, Romblon, Sibuyan.

143. Chibia palawanensis (Tweed.); Sharpe, Ibis, 1884, p. 318. Calamianes.

144. Chibia borneensis Sharpe; Sharpe, P. Z. S. 1879, p. 246. Tawi Tawi.

145. Buchanga palawanensis Whiteh.; Whiteh., Ibis, 1890, p. 47. Calamianes.

146. Oriolus chinensis (Linn.); Sharpe, Cat. B. Brit. Mus. xiii. p. 203 (1877). Tawi Tawi, Calamianes, Tablas, Romblon, Sibuyan, Masbate.

147. Munia oryzivora (Linn.); Sharpe, Cat. B. Brit. Mus. xiii. p. 328 (1890). Panay, Samar.

148. Munia jagori (Cab.); Sharpe, Cat. B. Brit. Mus. xiii. p. 337 (1890). Basilan, Tawi Tawi, Calamianes, Tablas, Romblon, Sibu yan, Siquijor.

149. Munia cabanisi Sharpe; Sharpe, Cat. B. Brit. Mus. xiii. p. 353 (1890). Panay.

150. Uroloncha everetti (Tweed.); Sharpe, Cat. B. Brit. Mus. xiii. p. 363 (1890). Calamianes, Tawi Tawi*, Tablas, Romblon, Sibuyan.

151. Passer montanus (Linn.); Sharpe, Cat. B. Brit. Mus. xii. p. 301, (1888). Cebu.

152. Motacilla ocularis Swinh.; Sharpe, Cat. B. Brit. Mus. x. p. 471 (1885). Palawan.

153. Motacilla melanope Pall.; Sharpe, Cat. B. Brit. Mus. x. p. 498 (1885). Sulu, Tawi Tawi, Tablas, Romblon, Sibuyan, Masbate.

154. Motacilla flava Linn.; Sharpe, Cat. B. Brit. Mus. x. p. 516 pl. vi. figs. 3-5 (1885). Negros.

155. Anthus gustavi Swinh.; Sharpe, Cat. B. Brit. Mus. x. p. 613 (1885). Tawi Tawi, Sibuyan, Romblon, Masbate.

156. Anthus rufulus Vieill.; Sharpe, Cat. B. Brit. Mus. x. p. 574 1883) Calam ianes, Tablas, Romblon, Sibuyan, Masbate.

157. Climacteris mystacalis (Temm.); Gadow, Cat. B. Brit Mus. viii. p. 339 (1883). Negros, Masbate, Samar.

158. Dendrophila oenochlamys Sharpe; Gadow, Cat. B. Brit. Mus. viii. p. 359 (1883). Samar.

159. Aethopyga magnifica Sharpe; Gadow. Cat. B. Brit. Mus. ix. p. 24 (1884). Panay, Tablas, Sibuyan.

160. Aethopyga shelleyi Sharpe; Gadow, Cat. B. Brit. Mus. ix. p. 29 (1884). Calamianes.

161. Cinnyris sperata (Linn.); Gadow, Cat. B. Brit. Mus. ix. p. 63 (1884). Calamianes, Tablas, Romblon, Sibuyan, Panay, Siquijor, Cebu.

162. Cinnyris juliae (Tweed.); Gadow, Cat. B. Brit. Mus. ix. p. 64 (1884). Tawi Tawi.*

163. Cinnyris jugularis (Linn.); Gadow, Cat. B. Brit. Mus. ix. p. 84 (1884). Tawi Tawi, Tablas, Romblon, Sibuyan, Masbate, Siquijor.

164. Cinnyris aurora (Tweed.); Gadow, Cat. B. Brit. Mus. ix. p. 88 (1884). Calamianes.

165. Cinnyris guimarasensis Steere; Steere, Ibis. p. 315. 1891, Panay.

166. Anthothreptes chlorigaster Sharpe; Gadow, Cat. B. Brit. Mus. ix. p. 123 (1884). Tablas, Sibuyan, Romblon, Panay, Tawi Tawi.*

167. Anthothreptes griseigularis Tweed.; Gadow, Cat. B. Brit. Mus. ix. p. 126 (1884). Samar.

168. Dicaeum rubriventer Less.; Sharpe, Cat. B. Brit. Mus. ix. p. 36 (1885). Samar, Masbate.

169. Dicaeum mindanense Tweed.; Sharpe, Cat. B. Brit. Mus. x. p. 37 (1885). Basilan, Sulu, Tawi Tawi.

170. Dicaeum pygaeum (Kittl.); Sharpe, Cat. B. Brit. Mus. x. p. 43 (1885). Samar, Siquijor, Masbate, Sibuyan.

171. Prionochilus johnannae Sharpe; Sharpe, Ibis. 1888, p. 201, pl. iv. fig. 1. Calamianes.

172. Zosterops everetti Tweed.; Gadow, Cat. B. Brit. Mus. ix. p. 163 (1884). Tawi Tawi*, Sulu.

173. Zosterops nigrorum Tweed.; Gadow, Cat. B. Brit. Mus. ix. p. 186 (1884). Masbate.

174. Parus elegans Less.; Gadow, Cat B. Brit. Mus. viii. p 22 (1883). Mindoro, Tawi Tawi.

175. Hyloterpe philippinensis Wald.; Walden, Trans. Zool. Soc. ix. p. 179, pl. xxxi. fig. 2. (1875). Mindanao.

176. Hyloterpe homeyeri Blas.; Blas. J. f. O. 1890, p. 143. Tawi Tawi, Mindanao.

177. Lanius lucionensis Linn.; Gadow, Cat. B. Brit. Mus. viii. p. 285 (1883). Sulu, Tawi Tawi, Calamianes, Tablas, Romblon, Sibuyan, Siquijor.

178. Lanius nasutus Scop; Walden, Trans. Zool. Soc ix. p. 169 (1875), Samar, Masbate, Calamianes.

179. Artamus leucogaster (Valenc.); Sharpe, Cat. B. Brit. Mus. xiii. p. 3 (1890). Tawi Tawi, Calamianes, Tablas, Romblon, Sibuyan, Masbate.

180. Phylloscopus borealis (Blas.); Seebohm, Cat. B. Brit. Mus. v. p. 40 (1881). Samar, Sulu, Tawi Tawi, Calamianes, Mindoro, Tablas, Sibuyan, Guimaras.

181. Cryptolopha olivacea (Moseley)=Abrornis olivacea Moseley; Ibis, 1891, p. 47. Mindanao, Sulu, Tawi Tawi.

182. Pratincola caprata (Linn.); Sharpe, Cat. B. Brit. Mus. iv. p. 195. (1879). Masbate, Siquijor.

183. Acrocephalus orientalis (Temm. & Schleg.); Seebohm, Cat. B. Brit. Mus. vii. p. 123 (1883). Mindoro.

184. Megalurus palustris Horsf.; Sharpe, Cat. B. Brit. Mus. vii. p. 125 (1883). Masbate.

185. Megalurus ruficeps Tweed.; Sharpe, Cat. B. Brit. Mus. vii. p. 224 (1883). Calamianes.

186. Orthotomus castaneiceps, Wald.; Sharpe. Cat B. Brit. Mus. vii. p. 223 (1883). Masbate.

187. Orthotomus ruficeps (Less.); Sharpe, Cat. B. Brit. Mus. vii p. 224. Calamianes.

188. Cisticola cisticola (Temm.); Sharpe, Cat. B. Brit. Mus. vii. p. 259 (1883). Mindanao.

189. Cisticola exilis (Vig. & Horsf.); Sharpe, Cat. B. Brit. Mus. vii. p. 269 (1883). Samar, Tablas, Sibuyan, Panay, Cebu, Siquijor, Masbate, Mindoro, Calamianes.

190. Geocichla interpres (Temm.); Seebohm, Cat. B. Brit. Mus. v. p. 166 (1881). Tawi Tawi.

191. Monticola solitaria (P. L. S. Müll.); Seebohm, Cat. B. Brit. Mus. v. p. 329 (1881). Sulu, Tawi Tawi, Calamianes, Tablas, Romblon, Sibuyan, Masbate, Siquijor.

192. Erithacus calliope (Pall.); Seebohm, Cat. B. Brit. Mus. v. p. 305 (1881). Masbate.

193. Copsychus mindanensis (Gm.); Sharpe, Cat. B. Brit. Mus. vii. p. 60 (1883). Tawi Tawi, Mindoro, Tablas, Sibuyan, Masbate, Siquijor.

194. Cittocincla nigra Sharpe; Sharpe, Cat. B. Brit. Mus. vii. p. 90 (1883). Calamianes.

195. Mixornis plateni Blas; Blas., J. f. O. 1890, pp. 145, 147. Samar.

196. Macronus kettlewelli Guillem.; Guillem., P. Z. S. 1885, p. 262, pl. xviii. fig. 2. Tawi Tawi.*

197. Chloropsis palawanensis Sharpe; Sharpe, Cat. B. Brit. Mus. vi. p. 33 (1881). Calamianes.

198. Iole haynaldi (Blas.); Sharpe, Ibis, 1894, p. 253. Tawi Tawi.

199. Poliolophus urostictus (Salvad.); Sharpe, Cat. B. Brit. Mus. vi. p. 79, pl. v (1881). Samar.

200. Criniger frater Sharpe; Sharpe, Cat. B. Brit. Mus. vi. p. 79, pl. v. (1881). Calamianes.

201. Pycnonotus goiavier (Scop); Sharpe, Cat. B. Brit. Mus. vi. p. 141 (1881). Samar, Tablas, Panay, Masbate.

202. Pycnonotus cinereifrons (Tweed.); Sharpe, Cat. B. Brit. Mus. vi. p. 153 (1881). Calamianes.

203. Irena melanochlamys Sharpe; Sharpe, Cat. B. Brit. Mus. iii. p. 266 (1877). Mindanao.

204. Irena tweeddalii Sharpe; Sharpe, Cat. B. Brit. Mus. iii. p. 268 (1877). Calamianes.

205. Artamides sumatrensis (S. Müll.); Sharpe, Cat. B. Brit. Mus. iv. p. 12 (1879). Calamianes.

206. Artamides guillemardi (Salvad.); Salvad., Ibis, 1886, p. 154. Tawi Tawi.

207. Artamides mindorensis Steere; Steere, List B. & M. Steere Exped. p. 14 (1890). Tablas.

208. Edoliisoma everetti Sharpe; Sharpe, Ibis, 1894, p. 251. Tawi Tawi, Sulu.

209. Pericrocotus leytensis Steere; Steere, List B. & M. Steere Exped. p. 15 (1890). Samar.

210. Lalage minor (Steere);—Pseudolalage minor, Steere, list B. & M. Steere Exped. p. 15 (1890). Samar.

211. Lalage terat (Bodd.); Sharpe, Cat. B. Brit. Mus. iv. p. 95 (1879). Calamianes, Tablas, Romblon, Sibuyan, Siquijor.

212. Muscicapa griseisticta (Swinh.); Sharpe, Cat. B. Brit. Mus. iv. p. 153 (1879). Siquijor, Masbate, Mindoro, Calamianes, Tawi Tawi.

213. Muscicapula mindanensis Blas., J. f. O. 1890, p. 147. Basilan.

214. Hypothymis azurea (Bodd.); Sharpe, Cat. B. Brit. Mus. iv. p. 274 (1879). Tawi Tawi, Calamianes, Tablas, Romblon, Sibuyan, Masbate, Siquijor.

215. Cyanomyias coelestis (Tweed.); Sharpe, Cat. B. Brit. Mus. iv. p. 278 (1878). Mindanao, Sibuyan.

216. Rhipidura nigritorquis Vigors; Sharpe, Cat. B. Brit. Mus. iv. p. 334 (1879). Samar, Tawi Tawi, Calamianes, Mindoro, Tablas, Romblon, Sibuyan, Masbate, Siquijor.

217. Zeocephus rufus Gray; Sharpe, Cat, B. Brit. Mus. iv. p. 343 (1879). Tablas, Romblon, Sibuyan, Cebu.

218. Zeocephus cyanescens Sharpe; Sharpe, Cat. B. Brit. Mus. iv. p. 343 (1879). Calamianes.

219. Zeocephus cinnamomeus Sharpe: Sharpe, Cat. B. Brit. Mus. iv. p. 343 (1879). Tawi Tawi.

220. Culicicapa panayensis, Sharpe; Sharpe, Cat. B. Brit. Mus. iv. p. 450 (1879). Tawi Tawi, Tablas, Romblon, Sibuyan, Guimaras. Masbate, Siquijor.

221. Siphia philippinensis Sharpe; Sharpe, Cat. B. Brit. Mus. iv. p. 371 (1879). Tawi Tawi, Tablas, Romblon, Sibuyan, Guimaras, Masbate, Siquijor.

222. Siphia lemprieri Sharpe; Sharpe, Ibis, 1884, p. 319. Calamianes.

223. Hirundo gutturalis Scop.; Sharpe, Cat. B. Brit. Mus. x. p. 134 (1885). Mindoro, Basilan, Sulu, Palawan, Siquijor.

224. Hirundo javanica Sparrm.; Sharpe, Cat. B. Brit. Mus. x.
p. 142 (1885). Luzon, Tawi Tawi, Calamines, Panay, Guimaras,
Masbate, Siquijor.

225. Pitta erythrogaster Temm.; Sclater, Cat. B. Brit. Mus.
xiv. p. 432 (1888). Tawi Tawi, Tablas, Sibuyan, Romblon, Panay,
Masbate, Cebu, Siquijor.

226. Pitta atricapilla Less.; Sclater, Cat. B. Brit. Mus. xiv. p.
438 (1888). Tawi Tawi, Calamianes, Mindoro, Tablas, Cebu.

The following known species not previously recorded from the
Philippines were found there by us:

1. Caprimulgus jotaka Temm. & Schleg.; Hartert, Cat. B. Brit.
Mus. xvi. p. 552 (1892). Palawan.

2. Prionochilus modestus Hume.; Hume, Str. F. p. 298, 1875.

We have therefore added 7 species to the known bird fauna of
Luzon, 7 to that of Palawan, 25 to that of Sulu, 15 to that of Basilan,
13 to that of Mindanao, 38 to that of Panay, 24 to that of Negros, 9
to that of Guimaras, 52 to that of Siquijor, 40 to that of Cebu, 37 to
that of Samar, 67 to that of Masbate, and 40 to that of Mindoro.

In Tawi Tawi we discovered 92 species, of which 19 were after-
ward found by Mr. Everett's collectors also, as well as three species
not obtained by us, making the whole number of species at present
known from Tawi Tawi 95.

In Romblon we discovered 44 species, in Sibuyan 65, and in Tablas
66.   The curious fauna of these three islands will be fully treated
of later.

In the Calamianes Islands (Culion and Busuanga) we discovered
73 species of birds, all of them identical with species found in Palawan. Most of the mammals found in the Calamianes Islands were
also well known Palawan forms. Deer are, however, very abundant
in the Calamianes, and we found one mammal in Busuanga which is
probably new.

IV.

ADDITIONAL NOTES ON PREVIOUSLY DESCRIBED SPECIES OF BIRDS.

Polyplectron napoleonis Less.

Polyplectrum napoleonis Less.   Traite d'Orn. pp. 487, 650 (1831).
Polyplectron enphanum Tem. Pl. Col. v. pl. 48 (No. 540) (1831).
Polyplectron emphanes Sclat. List of Phas. p. 12 (1854); Id. Ibis, 1878, p. 623.
Polyplectron napoleonis Tweed. (Nec. Less.) P. Z. S. 1878, p. 792; Sharpe, Ibis, 1888, p. 203; Blasius, Ornis, 1888, p. 317; Everett, Birds of Borneo, p. 198 (1889); Whitehead, Ibis, 1890, p. 57.
Polyplectron nehrkornae Blas. Mitth. orn. Ver. Wien, 1891, p. 1; Id. J. f. O. 1891. p. 10.

We have some information to offer concerning the habitat of the much discussed Polyplectron napoleonis. Mr. Ogilvie Grant states in the Catalogue of Birds, vol. xxii. p. 361, that the male of P. napoleonis is "exactly similar to the male of P. nehrkornae, but the white superciliary stripes are wide and strongly marked and confluent on the nape. Total length 18.5 inches, wing, 7.3, tail, 7.8, tarsus, 2.5. Habitat unknown. (?) Luzon, Philippine Islands."

While in Palawan we were so fortunate as to secure a series of eleven fully adult males of the Polyplectron inhabiting that island.

Of these, *two have not the slightest trace* of superciliary stripes, while a third has only four small white feathers on one side. In each of the above there are a few white feathers on the nape. Three of our specimens perfectly agree with the description of typical P. nehrkornae. Three specimens have *broad* superciliary stripes *almost* confluent on the nape, and in one bird the superciliary stripes, which begin between eye and nostril, are very broad, widening steadily towards the nape WHERE THEY ARE FULLY CONFLUENT. An examination of young birds of which we have a good series shows that the width and extent of the superciliary lines is independent of age. We therefore feel perfectly satisfied that P. Napoleonis and P. nehrkornae are identical since the width of the white superciliary stripes is an uncertain quantity, subject to wide individual variation, and may even be absent.

As P. napoleonis is easily trapped, we feel that the presumptive evidence against its occurrence in Luzon is very strong. The natives would certainly know of its presence were it found there and we venture to prophesy that Mr. Whitehead will search that island in vain for it.

The bill of P. napoleonis is black tipped with pale horn color. The legs, feet and nails are brown. The eyes chocolate brown. One of our birds has but one spur on the left leg.

In young birds the ocelli are at first grey with black centers. The irridescent blue-green color appears first in the middle of the ocellus and gradually spreads outward.

P. napoleonis is extremely shy. All our specimens were snared, though Mr. Bourns caught a glimpse of a female on one occasion. Our males average as follows: Length, 20.43. Wing, 7.10. Tail, 8.76. Culmen, .93. Tarsus, 2.40. The females are somewhat smaller: Seven specimens average 16.52 in length. Wing, 6.53. Tail, 5.94. Culmen, .87. Tarsus, 2.15.

Ardea jugularis of Dr. Steere's list is Demiegretta sacra, (Gm.).

Circus philippinensis Steere.

Circus philippinensis Steere, List B. & M. Steere Exped. p. 7 (1890).

Although Gurney, Everett and others doubt the validity of Dr. Steere's C. philippinensis and the presumptive evidence against the existence in the Philippines of a peculiar species of this genus would seem strong, the single female of our collecting most nearly agrees with Dr. Steere's description and we accordingly provisionally adopt his title.

Spilornis holospilus (Vig.).

Spilornis holospilus (Vig); Sharpe, Cat. B. Brit. Mus. 1. p. 293 (1874). Spilornis panayensis Steere, List B. & M. Steere Exped. p. 7 (1890).

Dr. Steere has attempted to separate the representatives of this genus from the central Philippines under the name S. panayensis, on the ground that they are smaller and lighter in color than is S. holospilus. We find that both light and dark birds occur throughout the range of the species in the islands. We have very dark and richly colored birds from the central islands, *but we do not find any constant difference in size* between them and birds from other parts of the group. We think that S. panayesis was founded on differences due to change of season and to individual variation and believe that the species is not a valid one.

Ninox japonica (T. & S.).

Ninox japonica (T. & S.); Sharpe. P. Z. S. 1879, p. 325.

There is no doubt as to the identity of the birds in question, which are from Cebu. Three males measure 12.62 inches in length. Wing, 8.98. Tail, 5.37. Culmen, .71. Tarsus, 1.27.

Ceyx euerythra Sharpe.

Ceyx euerythra Sharpe, Cat. B. Brit. Mus. Vol. xvii. p 179 (1892).

It is hard to make out what Dr. Sharpe means by his description (Cat. B. xvii. p. 179) "entirely red with a wash of beautiful lilac on the upper surface; greater wing-coverts and innermost secondaries tipped with rufous," but we take it that he means *upper surface* entirely red. We have fourteen specimens from Tawi Tawi, Palawan and the Calamianes islands and they show some interesting plumage changes which were for some time a puzzle to us.

An adult pair in high plumage from the Calamianes have entire upper surface of body red, washed with lilac, most heavily on head and rump. Wing-coverts and scapulars like back. Secondaries rufous with broad black shaft stripes. Primaries black, the first rufous for entire length of outer web and most of the others showing a small amount of rufous at their tips. Under surface of primaries washed with rufous on inner webs. Chin and throat white faintly tinged with lemon yellow. Remainder of under surface deep golden yellow, darkest on sides of breast and flanks. A yellowish white patch behind ear. Under tail-coverts golden yellow tipped with rufous. Tail uniform bright rufous above and below. Under wing-coverts and axillaries golden yellow. Bend of wing rufous.

This plumage we take to be typical for fully adult birds in fine feather.

Two females from Palawan agree with this description except that in one the secondaries show rufous only on under surface of inner webs, and that the primaries show no rufous except on basal half of outer web of first. The under wing coverts, axillaries and bend of wing are light rufous instead of yellow.

Dr. Sharpe (Ibis. '94, p. 246) reports an adult male of this species from Bongao and a female from Tawi Tawi. We note certain differences shown by our Tawi Tawi specimens, of which we have twelve. Of these four agree with the Calamianes birds except that the scapulars show more or less black at their bases and that the secondaries show more of black. Among the remaining specimens, however, there are some curious variations.

First it is to be noted that in three fully adult birds beginning to moult the under surface is pale dirty yellowish, the throat white.

In two of the birds a few scattered yellow feathers are appearing in the white of the throat. This then is the worn out plumage of old birds.

A male with rich yellow under surface and white throat has some of the scapulars entirely black, tipped with blue, the remainder

being tipped with lilac.   Some of the wing-coverts are black tipped with blue.   No rufous on primaries except on outer web of first.

Another bird has chin and throat pure white, the breast mottled
with golden yellow and light cinnamon rufous.   Feathers of abdomen nearly white, tips washed with rufous.   Under wing-coverts
and axillaries cinnamon rufous.   A little more black in the scapulars than the preceding.   Tail with broad black shaft stripes on
apical half of under surface of feathers.

Another specimen has chin and throat pure white.   Sides of face,
breast, flanks, under wing-coverts and axillaries cinnamon rufous,
deepest on the breast.   Abdomen nearly white.   A few golden yellow feathers appearing on breast, flanks and abdomen.   Scapulars,
except a few of the smallest, black quite broadly tipped with blue,
tail with tips of all its feathers black.

Finally, a single specimen has under surface as in preceeding
except that yellow feathers have not begun to appear.   Scapulars
and inner third of inner secondaries black, the former tipped with
blue, the latter with rufous washed with lilac.   Tail with apical
two-thirds of feathers black washed with rufous on edges of webs.
The bill of this last bird shows signs of immaturity, being blackish
toward the tip instead of clear scarlet.

We were at first greatly puzzled by these birds, as the black
scapulars with their blue tips form a striking marking and with a
single exception the bills of our specimens showed no sign of immaturity.   After carefully examining the whole series, however, we
are convinced that the cinnamon rufous under surface. tail-feathers
tipped with black and black scapulars tipped with blue are themselves signs of immaturity, the black gradually disappearing with
age, and yellow feathers appearing on the under surface
until the plumage first described by us is reached.   This finally
becomes worn and soiled giving the dirty yellowish under plumage
already noted.   Our Tawi Tawi birds were shot late in October and
early in November, Palawan birds in December, and Calamianes
birds in January and February.

Ceyx melanura Kaup.

Ceyx melanura Kaup.; Sharpe, Cat. B. Brit. Mus. xvii. p. 180 (1892).
Ceyx samarensis Steere, L'st B. and M. Steere Exped. p. 10 (1890).

We obtained a fine series of specimens in Samar which agree in
every detail with the description of C. melanura.   The specimens
secured by us were found among the hills in deep forest and invariably away from water.

Ceyx mindanensis Steere

Ceyx mindanensis, List B. and M. Steere Exped. p. 10 (1890).
Ceyx basilanica Steere, List B. and M. Steere Exped. p. 10 (1890). Sharpe,
   Cat. B. Brit. Mus. vol. xvii. p. 181 (1892).
Ceyx, platenæ Blas., J. f. 10. 1890. p. 141.

With a very large series of specimens from Mindanao and Basilan
at our disposal we are unable to detect the slightest difference
between the birds from the two islands and we therefore unite them
under the name C. mindanensis, as the Mindanao birds were obtained
and described first.   Frequents forest or low second growth away
from water.

Ceyx bournsi Steere.

Ceyx bournsi Steere, List B. & M. Steere Exped, p. 10 (1890); Sharpe, Cat.
   B. Brit. Mus. p. 185, vol. xvii (1892).
Ceyx malamaui Steere, List B. & M. Steere Exped. p. 10 (1890); Sharpe,
   Cat. B. Brit. Mus. p. 184, vol. xvii (1892).
Ceyx suluensis Blas., J. f. O. p. 141 (1890).
Ceyx margarethae Blas., J. f. O. 1890, p. 141.

It will be noticed, doubtless with some surprise that we have
here united several apparently well marked species.   We can only
request those who are inclined to doubt the reasonableness of our
action to postpone their final decision until they have inspected our
series of specimens.   Our conclusions are based on a series of
sixty-six specimens from Tablas, Romblon, Sibuyan, Negros, Siquijor, Cebu, Basilan, Sulu and Tawi Tawi.

We find that we must either multiply the number of small blue
woods Ceyces from the Philippines indefinitely or reduce the
above mentioned species to one.   It would be an almost endless
task to describe the different phases of plumage shown and we will
only say that we have a practically unbroken series between a bird
with a magnificent deep blue upper surface and a bird with a fine
silvery white upper surface *which has not a blue feather on it.*

In the latter specimens the white occupies exactly the position of
the blue in the specimens first mentioned.

Our series shows that these extraordinary differences of color are
independent of sex, age or locality, some young birds are very
light, others very dark.   In one case where parent and offspring
were killed at one discharge of the gun they exhibited marked differences in color.

The *amount* of blue or white is, however, dependent on age to
some extent, the young birds always showing much more black on
the upper surface than do adults.   Bill, legs, feet and nails
are bright scarlet in adults.   In the young the bill is at first black
tipped with pale horn and the legs and feet are pale flesh color.

We retain the name C. bournsi because some seventy-five per cent of our specimens answer fairly well the description of that species as given by Dr. Sharpe, Cat. B. xvii. p. 185. When this type is once departed from, however, the variations are interminable and with a smaller series of specimens at our disposal we should certainly have fallen into hopeless confusion. We note, also, great variability in the color of the under surface, some specimens being very much darker than others.

Ceyx bournsi is a strictly woods form and its shy habits doubtless explain its having been so generally missed by collectors.

Ceyx cyanipectus Lafr.

Ceyx cyanipectus Lafr.; Sharpe, Cat. B. Brit. Mus. xvii. p. 185 (1892).
Ceyx steerii Sharpe, Cat. B. Brit. Mus. xvii. p. 187 (1892).

It is with regret that we find ourselves compelled to differ from so eminent an authority as Dr. Sharpe, but an examination of a large series of specimens has convinced us that the Luzon and Mindoro birds are identical. C. cyanipectus is extremely common along the fresh water steams in the interior of Mindoro. On one occasion we shot thirteen specimens in a single day.

Dr. Sharpe records one female only from Mindoro in the British Museum collection. He separates the Mindoro bird on accout of "the dull reddish color of the under surface," *but Luzon birds in fine plumage* SHOW THIS SAME COLOR. The plate in Ibis, 1884, is extremely poor and gives an entirely erroneous idea of the color of the under parts of the Luzon bird. The plate in Dr. Sharpe's monograph of the Alcedinidae is much nearer the truth.

We are forced to the conclusion that Dr. Sharpe has fallen into error on account of insufficient material, though the trouble seems to be rather with the Luzon specimens than with the one from Mindoro. If the plate in Ibis correctly represents the specimens obtained by Mr. Maitland-Herriot they must have been in extremely poor plumage. The other British Museum specimens from Luzon are apparently all old and may be faded.

We found C. cyanipectus along the banks of fresh water streams in Masbate and Sibuyan as well as in Mindoro and Luzon.

In Sibuyan a single specimen was seen in a mangrove swamp. We never met with it away from water.

SPECIES PREVIOUSLY DESCRIBED. 49

Anthracoceros montani (Oust.).

Buceros montani Oust. Bull. Hebd. Assoc. Scien. de Fr. p. 206 1880.
Anthracoceros montani Grant, Cat. B. Brit. Mus. xvii. p. 370 (1892).

We were so fortunate as to secure a series of fourteen specimens of this rare horn-bill from Sulu and Tawi Tawi. The tail is pure white. All other parts black, the feathers of back and wings glossed with dark green. The bill in abult birds is coal black. In all of our adult males the iris was nearly white while in the females it was dark brown. Legs and feet dull leaden, nails black. Young birds have tip of bill white or pale horn.

Fairly common on the hills back of the town of Sulu and very abundant in Tawi Tawi where it occurs in great flocks. It is a very wild bird, always difficult to approach. Its cry is the most peculiar bird note we have ever heard. It begins with a series of notes precisely like the "song" of a common hen magnified about fifty fold and ends with an indescribable combination of cackles and shrieks.

Three males average 28.12 inches in length. Wing, 11.52. Tail 9.51. Tarsus, 2.02. Seven females measure 26.73 inches in length. Wing, 10.85. Tail, 8.90. Tarsus, 1.94.

Collocalia francica (Gm.).

Collocalia francica (Gm.); Hartert, Cat. B. Brit. Mus. xvi. p. 503 (1892).

Hartert (J. f. O. 1891, p. 302) seems strongly inclined to doubt the occurence in the Philippines of C. francica and suggests that the birds so identified by Dr. Steere may have been C. marginata. Were there no other distinction the great difference in the size of the two species would make such a blunder impossible.

We obtained a series of specimens in Culion and Panay which answer the description of C. francica as given by Hartert (Cat. B. Brit. Mus. xvi. p. 503) in every detail, having the band of smoky white feathers with distinct black shaft stripes well defined on the rump. Hartert states that the length of C. francica is "over four inches." Our specimens measure 4.62 inches in length. Wing, 4.54. Tail, 1.97. Culmen, .21. Tarsus, .37.

Iris dark brown. Legs and feet light brown. Bill black.

Surniculus velutinus Sharpe.

Surniculus velutinus Sharpe; Shelley, Cat. B. Brit. Mus. xix. p. 230 (1891).

Abundant in Basilan. A young bird from this island, two thirds grown, is light rusty brown in color, lightest on under surface. The crown and nape show metallic blue black feathers. One of the scapulars, many feathers of rump and all of tail same color, mostly tipped with rusty brown. Wing-coverts rusty brown with faint dark shaft markings appearing. Upper surface of wing black, faintly metallic, all the feathers edged with rusty brown. Under surface of body uniform light brown. Wing-coverts white strongly washed with brown at tips. White spot on inner webs of primaries appearing. Under surface of tail dull metallic blue, each feather with several spots of white.

A bird nearly grown shows numerous brown feathers on head and back. Primaries washed with rusty brown, primary coverts uniform brown. Throat, chin and upper breast nearly black. A third has general color of adult but some feathers of head, nape, primary coverts, breast and abdomen as well as tips of some of the secondaries are washed with rusty brown.

Bolbopsittacus intermedius Salvad.

Bolbsittacus intermedius Salvad., Cat. B. Brit. Mus. xx. p. 505, pl. xiii. (1891).

We are happy to be able to establish the habitat of Count Salvadori's B. intermedius. Four specimens were secured near Catbologan, Samar. An adult male agrees perfectly with Salvadori's description and excellent figure and there is no room for doubt as to the identity of the Samar birds.

The female has never been described. It differs from the male in having the blue of the head confined to the throat, the cheeks being light green. Around eye a ring of green lighter than that of crown. The blue collar is replaced by an indistinct collar of faint orange yellow. Rump only slightly lighter than back and green, not yellow as in male. Under surface slightly lighter and more yellowish. Thighs green instead of yellow.

A young male is like the female but has less blue on sides of throat.

Loriculus worcesteri Steere.

Loriculus worcesteri Steere, List B. & M. Steere Exped. p. 8 (1890).
Loriculus apicalis Salvad. (Part. Samar & Leyte only), Cat. B. Brit. Mus. xx. p. 528 (1891).

Count Salvadori identifies L. worcesteri with L apicalis. He records a single specimen collected by Mr. E. L. Moseley of the Steere Expedition, as being in the British Museum collection. We are decidedly of the opinion that Dr. Steere's determination will hold good and that Salvadori will agree with us when he has a larger series of specimens at his disposal. The Marquis of Tweeddale noted certain differences between the birds of this genus collected by Mr. Everett in southern Leyte and typical L. apicalis from Mindanao and provisionally identified them with the latter species pending far'her investigation.

The differences between the Samar and Mindanao birds are: *First*, that the red mark on crown of Samar birds is *distinctly narrower* than in those from Mindanao and *tapers sharply to a point* on the nape instead of spreading out and *ending broadly.*

Second, the feathers of the back are *barely tinged* with golden, not one of our specimens showing anything like the amount of color exhibited by Mindanao birds.

The Samar and Mindanao birds can be readily separated by the head markings alone and there is far more difference between them than between other species recognized by Salvadori, such, for instance, as Prioniturus discurus (Vieill.) and Prioniturus suluensis Salvad.

Xantholaema intermedia Shelley.

Xantholaema intermedia Shelley, Cat. B. Brit. Mus. xix. p. 97 (1891).

We note certain plumage difference between the Cebu, Negros and Tablas birds. The Negros birds best agree with Shelley's description. Tablas birds also agree with the description in having feathers of back without lighter edges, but they have the black stripe bordering red of throat washed with *olive green* instead of olive grey. The Cebu birds, which were obtained in June, have light edgings to feathers of upper surface, show more yellow in the spots above and below eye than do birds from Negros and Talbas, and have stripe between eye and throat almost pure black, sometimes faintly washed with olive grey. They also differ, in that the black of hind crown and nape is much less heavily washed with olive grey.

The Philippine representatives of the Genus Iyngipicus.

Much confusion still exists as to the Philippine representatives of this genus, witness Hargitt's recording I. maculatus (Scop.) from Luzon (!). The type of this species was obtained by Sonnerat at Antique ("Antigua") in north Panay.

Lord Walden (P. Z. S. ix. p. 148, 1875) united the Luzon, Panay and Mindanao birds on the *supposition* that the three islands possessed but one species. This supposition he afterwards had occasion to modify when he obtained some material on which to work.

The Luzon birds collected by Everett were identified by Walden as I. maculatus and the Mindanao-Basilan species as I. validirostris. In Ibis, 1881, p. 597, Hargitt very properly separates the Mindanao-Basilan form under the name I. fulvifasciatus. In J. f. O. 1882, p. 227, Kutter describes a single specimen from Guimaras under the names "Yungipicus maculatus Scop." and "Baepipo validirostris Cab." Kutter's example was a poor one, as he expressly states, and having no material for comparison he unhesitatingly united the Guimaras and Luzon birds.

Dr. J. B. Steere, the first naturalist before whom a series of specimens from all these localities ever lay, saw at once the differences between the Luzon and Panay birds, which unfortunately he seems not to have thought worth pointing out. and rightly retained the name "maculatus" for the Panay species. The Luzon birds he called "validirostris," with apparent reason as they most certainly are *not* maculatus and the birds described by Blyth may well have come from Luzon. The Mindanao-Basilan species he re-described under the name "Yungipicus basilanica," overlooking Hargitt's name and description entirely.

We note the following differences between the Luzon-Marinduque-Mindoro birds and those from Panay, Guimaras, Negros and Cebu. I. validirostris, the first mentioned species, has "the upper parts, together with wings, tail, also their coverts brownish black." I. maculatus has these parts *very light rusty brown* and the wing-coverts and quills are spotted with *buffy* white, not *pure* white as in I. validirostris. Bars on tail buffy white. Shafts of quills rusty brown. Occiput, nape and hind-neck *rusty brown like back*, the black of I. validirostris being entirely lacking. A *broad* scarlet stripe on side of occiput, the stripes of opposite sides being nearly confluent on nape. No black tips on nasal plumes. Auricular stripe rusty brown, *not brownish black.* The fulvescent wash on under surface of I. validirostris *usually entirely lacking* and *always* very faint. Ill-defined rusty brown stripes on under surface in place of sharp black markings of I. validirostris. I. maculatus is also lighter on rump than is I. validirostris, the feathers being faintly spotted with lighter color and *not barred* as in I. validirostris. The adult female of I. maculatus is like the male but lacks the scarlet markings on head.

The other species of the genus found in the Philippine islands with the possible exception of that from Samar and Leyte are so well marked that no possibility of confusion exists.

Chrysocolaptes rufopunctatus Harg.

Chrysocolaptes rufopunctatus Harg. Ibis, 1889, p. 231.
Chrysocolaptes samarensis Steere, List B. & M. Steere Exped. p. 8 (1890).

Another case in which Dr. Steere re-described a species already known.

Thriponax philippinensis Steere.

Thriponax philippinensis Steere, List B. & M. Steere Exped. p. 8 (1890); Id. Ibis, 1891, p. 305.

Dr. Steere's description of this species is so brief that we venture to add to it somewhat.

Adult male in breeding plumage. Forehead and crown to nape brilliant scarlet. Elongated feathers of crown and nape with yellowish white bases and shaft stripes of same color extending half their length. A scarlet stripe from base of lower mandible to ear. Lores, stripe under eye, auricular region, chin, throat and sides of neck black, some of the feathers tipped with scarlet, others with creamy white. Feathers of thighs buffy white with broad central spots of black. Lower breast, abdomen, flanks, under wing-coverts, axillaries, basal fifth of inner webs of secondaries and a narrow stripe on rump creamy white. All other parts black, tips of primaries and tail-feathers rusty. Feathers of fore-breast uniform black except a few of those immediately bordering the white of the breast which are tipped with that color. Many feathers of hind-neck and interscapulars broadly tipped with scarlet but the latter markings, as well as scarlet and white tips on feathers of sides of face and throat, are very variable.

Adult female. Like the male, but has forehead and crown pure black and lacks the scarlet cheek patch. Few of the feathers of head and neck are tipped with white and none are tipped with scarlet.

Sarcophanops steerii Sharpe.

Sarcophanops steerii Sharpe; Sclater, Cat. B. Brit. Mus. xlv. p. 482 (1888).

There has been some difference of opinion between Dr. Steere and Mr. A. Everett as to the color of the eyes of this interesting species. Both were right and there was abundant room for still more divergence of opinion. The eyes of S. steerii are golden yellow, bright green or a beautiful blue according to the way the light strikes them.

The locality "Mindoro" given for this species in Dr. Steere's list is a misprint for *Mindanao*. No representative of the genus was found by the Steere Expedition in Mindoro.

The young show some interesting plumage changes. An immature male has the under surface white, some of the feathers tipped
with pale lilac. Chin black but throat white, a few black feathers
just appearing. Head as in adult but white nuchal collar much
narrower. Back and wing-coverts washed with olive green, the
wing-bar being ill-defined and paler than in adult. Rump and tail
as in adult. Bill as in adult except center of upper mandible
which is black.

Another young male, slightly older, has chin and throat black,
the feathers narrowly tipped with white and shows more lilac
on breast. Crown, nape and back washed with olive green, purple
appearing on one or two feathers of forehead. Bill pure black.

A young female is like the first young male described but without lilac on under surface.

Sarcophanops samarensis Steere.

Sarcophanops samarensis Steere, List B. & M. Steere Exped. p. 23 (1890); id.
    Ibis, 1891, p. 316.

Adult male: Much smaller than S. steerii. Head, back and
scapulars purple somewhat mottled with brown. White nuchal
collar very narrow and ill-defined. Purple of back gradually
changing into brown on rump. Tips of scapulars black. Upper
tail-coverts and tail bright chestnut. Upper wing coverts black.
Tertiaries barred across both webs with pure white. Three secondaries with lilac spot on outer webs. Tips of secondaries and tertiaries black Primaries blackish brown. Chin, throat, sides of
face, ear-coverts and lores pure black. Breast, abdomen and flanks
lilac, deeper on upper breast, lighter on abdomen. Thighs black,
the feathers tipped with brown. Under tail-coverts light buff. Axillaries white. Under wing-coverts black. Bend of wing white.

Female like male except that the lilac of under surface is replaced
by white. Bill, legs, feet, nails and eyes exactly as in S. steerii
and the young show the same plumage changes as in that species.

Average measurements from five males: Length, 6.02 inches.
Wing, 3.06. Tail, 2.44. Culmen, .87. Tarsus, .80.

Ten males of S. steerii average 6.86 inches in length. Wing, 3.33.
Tail, 2.37. Culmen, .93. Tarsus, .86.

Oriolus samarensis Steere.

Oriolus samarensis Steere, List B. & M. Steere Exped. p. 17 (1890); id. Ibis,
    1891, p. 311.

Sexes alike. General cover above slightly lighter than in O.
steerii. Lores, chin, throat and upper breast uniform light ashy
grey lighter than in O. steerii. Ear-coverts and sides of neck like

back. Wing much yellower than in O. steerii, the yellow extending down to outer webs of primaries. Black markings of lower breast and abdomen narrower than in O. steerii. Axillaries and under wing-coverts, inner half of inner webs of secondaries and inner edges of basal half of primaries bright yellow. Central tail-feathers unmarked as in O. steerii.

Average measurements from five males. Length, 7.72 inches. Wing, 4.15. Tail, 2.85. Culmen, .95. Tarsus, .81.

Cinnyris guimarasensis Steere.

Cinnyris guimarasensis Steere, List B. & M. Steere Exped. p. 22 (1890); id. Ibis, 1891, p. 315.

It is to be regretted that Dr. Steere did not find some more appropriate name than "guimarasensis" for this beautiful species. Guimaras is, zoölogically speaking, a part of Panay, and the species in question was found in Panay by Mr. Worcester a few weeks after its discovery in Guimaras. Upon our return to the Philippine islands we obtained a fine series of specimens from the mountains of Panay. The female seems never to have been described. *Adult female:* Head and nape light olive green, becoming browner on back, wing-coverts and outer webs of secondaries. Upper tail-coverts like back. Tail black, webs of central pair of feathers washed with same color as back. Sides of face dark ashy grey, edges of feathers darker than centers. Chin light yellow. Entire throat grey faintly washed with yellow. Entire breast bright orange yellow paler on flanks, abdomen and under tail-coverts. Axillaries, under wing-coverts and inner webs of quills pure white.

Dicaeum trigonostigma and its Philippine allies.

It is well known that Dr. Sharpe identified a Dicaeum brought back by Dr. Steere in 1874 and supposed to have come from Negros, as D. trigonostigma. This identification has since been called in question by the Marquis of Tweeddale and others.

The specimen in question is still in existence in the Museum of the University of Michigan. There is not the slightest doubt that it is D. trigonostigma but in our opinion there is *very grave* doubt as to its *ever having come from Negros.* By some means labels seem to have become displaced on a number of the birds collected by Dr. Steere at this time (e. g. Parus elegans from Palawan). The specimen in question no longer bears Dr. Steere's original label and the Doctor

himself does not feel at all sure that it came from Negros. D. trigonostigma may, then, be safely excluded from the list of Philippine Dicaeidae, especially since the bird fauna of Negros is well
known, and no other collector has ever met with it there, whereas,
D. dorsale is very abundant in the island.

There are, however, nine species of Dicaeidae in the Philippines
which must, we think, be regarded as representative species of D. trigonostigma.   One of these, D. besti Steere, from Siquijor, has not
been very completely described and the female was unknown to
Dr. Steere.

Dicaeum besti is a well marked species discovered by ourselves in
1888.   It is apparently confined to the little island of Siquijor.

Adult male.   Above like D. cinereigulare but with a slightly
heavier yellow wash on rump.   *Chin bright yellow.*   Throat and upper breast bluish gray.

Adult female.   Entire upper surface and sides of head and neck
slaty grey washed with olive green most heavily on back and upper
tail-coverts.   Wing-coverts and secondaries edged with olive green.
Rest of upper surface of wing dark blackish brown, the primaries
edged with ashy grey.   Chin and throat light yellow.   Sides of
throat, breast and flanks ashy grey washed with yellow.   A stripe
of bright yellow beginning on breast and running down abdomen.
Under tail-coverts bright yellow.   Under wing-coverts and inner
webs of quills white faintly washed with yellow.   Axillaries light
yellow.

The males of the nine species may be characterized as follows:

D. xanthopygium has *yellow rump* and throat and *orange* breast.

D. intermedia has the *rump slaty blue* very faintly washed with olive
green.   Chin and throat *grey uniformly washed with yellow*.   Breast
pale orange.

D. sibuyanica has rump faintly washed with yellow, a *clear uniform blue grey throat* and a pale orange breast.   It is farther distinguished by its *size*, being the largest representative of the D. trigonostigma type yet discovered in the Philippines.

D. dorsale.   *Rump uniform with back* or very faintly tinged with
olive green.   Under surface *intense orange* usually paler on throat.

D. pallidior.   *Rump uniform with back.   Under surface paler than in
any other Philippine representative of the D. trigonostigma type.* Breast
only *very faintly* orange.

D. besti.   Male has rump washed with olive.   *Chin yellow, throat
grey,* breast *bright orange.*   The female also differs strikingly from
that of any other Philippine species having the throat much like
that of the male.

D. cinereigulare. *Chin and upper throat yellow. Sides of throat and lower throat grey washed* with yellow. Breast *brilliant orange* nearly equalling that of D. dorsale in intensity.
D. assimilis. Rump heavily washed with yellow. *Throat ashy* MUCH DARKER than in D. sibuyanica. Breast tinged with orange.
D. sibutense. *Throat uniform with head.* Lower back and rump slightly washed with olive.

Each of these species is, so far as we know, confined to a definite area so that no two overlap.

Prionochilus quadricolor Tweed.

Prionochilus quadricolor Tweed.; Sharpe, Cat. B. Brit. Mus. x. p. 70 (1885).

This fine species is not uncommon in the forest of Cebu but seems never to be found in open country, and as the small amount of forest remaining on the island is rapidly being cleared away we fear that P. quadricolor will become extinct before many years if, as seems at present possible, it is confined to the island of Cebu.

We are now able to furnish descriptions of the adult female and young.

Adult female. Above brown, head faintly washed with olive. Back, wing-coverts and outer webs of secondaries heavily washed with olive yellow. Rump almost entirely of the latter color. Entire under surface greyish white faintly tinged with olive yellow, lighter along center of breast and abdomen. Under wing-coverts, axillaries and inner webs of quills pure white.

Young male like female but lacks yellow wash on outer webs of secondaries, the wing and tail being black as in adult male.

Prionochilus modestus Hume.

Prionochilus modestus Hume. Str. F. p. 289 (1875).

A series of birds obtained by us in Palawan most nearly agree with the above species and are provisionally so identified pending comparison.

Zosterops everetti Tweed.

Zosterops everetti Tweed.; Gadow, Cat. B. Brit. Mus. ix. p. 163 (1884).
Zosterops basilanica Steere, List B. & M. Steere Exped. p. 21 (1890); id. Ibis, 1891, p. 314.

After examining a large series of birds from Cebu, Samar, Mindanao, Basilan, Sulu and Tawi Tawi we have come to the conclusion that there is no ground for separating the birds from the south. The specimens collected by the Steere Expedition were obtained in the month of November while those from Cebu, Leyte and Samar

were obtained in March and April, The differences enumerated by
Dr. Steere (1. c.) certainly existed between the Cebu and Basilan
birds but we failed to detect them in the birds from Samar.   Our
private collections contained typical Z. everetti from Samar and the
birds from that island were so identified by us.

On our present trip we again collected in Basilan in the autumn
and the same differences show themselves between the birds then
collected and those obtained in Cebu in July of the following year.
Our collections in Sulu and Tawi Tawi were made later than those
from Basilan and while many of the birds secured there are young,
or in poor plumage, we have typical Z. everetti from both localities. Dr. Sharpe also records Z. everetti as collected in Tawi Tawi
and Bongao by Everett but we do not know in what month.

We find no difference in either the breadth or depth of color of
the yellow stripe on under surface of the northern and southern
birds and think the slight difference in the amount of yellow about
the lores to be purely a matter of season, the amount of yellow
increasing as the breeding time approaches.

Cittocincla cebuensis Steere.

Cittocincla cebuensis Steere, List B. and M. Steere Exped. p. 20 (1890); Id.
Ibis. 1890. p. 314.

Apparently confined to the island of Cebu where it is very rare.

An immature female is slaty black above, tail dull black. Wing-
coverts brown tipped with distinct spots of rufous brown, these
spots forming two irregular bars. Quills fulvous brown. Primaries faintly washed on outer webs with rufous brown. Forehead
brownish. Lores, ring around eye, sides of face and chin light
rufous brown. Center of throat and upper breast slaty grey, a few
of the feathers still retaining brownish centers. Rest of under
surface slaty black washed with brown. Under tail-coverts black
with brown shaft stripes.

Mixornis plateni Blas.

Mixornis plateni Blas. J. f. O. pp. 145, 147, 1890.

A Mixornis apparently of this species, with which it agrees in
size, was found by us in Samar. But two specimens were secured.
One of these is immature. The head was broken from the other in
shipping and has been lost so that we are unable to identify the
Samar birds with absolute certainty, but they are either M. plateni
or a very closely allied species. Length of adult bird 4.25 inches.
Tarsus, .56. Tail, 1.85. Wing, 2.12. Iris with yellowish white inner
and red outer ring.

An immature male is olive brown above, the feathers of head and back with white shaft stripes. Tail fulvous brown the feathers edged with rusty brown. Wing coverts like back. Quills fulvous brown, their outer edges washed with light rufous brown. A line of nearly black feathers with white shafts over eye. Ear-coverts dark, with distinct white shaft lines. Feathers of upper throat blackish with broad white shaft markings. Lower throat and breast dark ashy, some of the feathers with ill-defined lighter shaft markings. Lower breast and abdomen nearly white, the feathers soft and fluffy. Flanks ashy grey. Under wing-coverts axillaries, inner webs of quills and under tail-coverts buffy white.

Cyanomyias helenae Steere.

Cyanomyais helenae Steere. List B. & M. Steere Exped. p. 16 (1890); id. Ibis, 1891, p. 311.

We obtained a good series of specimens of this fine species and are able to furnish a description of the *adult female*, which is dull verditer blue above, much brighter on head. Forehead and line over eye cobalt blue. Feathers of crest but slightly elongated. Tail brownish black washed with verditer blue. Shafts of feathers black. Wing coverts and outer edges of quils like back. Chin bluish grey. Cheeks, ear-coverts, throat and upper breast azure blue, brightest on cheeks. Abdomen and under tail-coverts white. Under surface of tail dark brown. Shafts white. Under wing-coverts and axillaries grey broadly tipped with white. Inner webs of quills washed with white. Bend of wing washed with verditer blue.

Four males measure 5.40 inches in length. Wing, 2.85. Tail, 2.90. Culmen, .61. Tarsus, .64. A female is larger measuring 6.06 inches in length. Wing, 2.86. Tail, 2.70. Culmen, .64. Tarsus, .67.

Orthotomus castaneiceps Wald.

Orthotomus castaneiceps Wald.; Sharpe, Cat. B. Brit. Mus. vii. p. 223 (1883).
Orthotomus panayensis Steere, List B. & M. Steere Exped. p. 20 (1890); Id. Ibis, 1891, p. 314.

Dr. Steere has attempted to separate the Panay tailor bird from that of Guimaras and Negros but after a most careful examination of a large series of specimens from Panay, Negros and Masbate we are compelled to say that there is not the slightest difference between the birds from the three islands. Their size is the same. The wash of olive green on the back, on which Dr. Steere relied to separate the Panay birds, is a variable character present in some birds, absent in others shot at the same season. It occurs in birds from Negros and Masbate as well as in those from Panay. The presumptive evidence against finding one species of Orthotomus in Panay and another in Guimaras is of course very strong. Guimaras is to all intents and purposes a part of Panay and there are no other known differences between the birds of the two islands. Masbate is a new locality for the species.

Iole philippinensis (Gm.).

Iole phillippensis (Gm.); Sharpe, Cat. B. Brit. Mus. vol. vi. p. 58 (1881).
Iole guimarasensis Steere, List B. & M. Steere Exped. p. 19 (1890); id. Ibis, 1891, p. 313.

Dr. Steere separates the Iole from Panay, Guimaras and Negros from the Luzon, Samar, Bohol, Cebu, Leyte and Mindanao birds. He states that I. guimarasensis has the "size and general coloring of I. rufigularis, with the light shaft streaks of I. philippinensis."

The latter character would not seem to be of especial value in differentiating it *from* I. philippinensis and we can find nothing in the size or color of our large series of specimens from the central Philippines to warrant us in separating them from typical I. philippinensis.

Dr. Steere mentions the very different note of the Cebu birds. We were unable to perceive the slightest difference in the notes of the birds in question and incline to the opinion that the doctor must have heard the note of I. monticola when he crossed over into Cebu.

V.

LIST OF MAMMALS COLLECTED.

Macacus philippinensis Geoffr.

Macacus philippinensis Geoffr.; Günther, P. Z. S. 1879, p. 74.

Occurs in every island visited by us. Tamed and carried everywhere by the natives.

Nycticebus tardigradus Linn.

Nycticebus tardigradus Linn.; Everett, Mammals of the Bornean Islands, p. 494 (1893).

Common in Bongao. Less common in the part of Tawi Tawi visited by us but occurs there. Called "shame face" by the Spaniards because of its curious habit of hiding its face.

Tarsius spectrum Pallas.

Tarsius spectrum Pall.; Everett, P. Z. S. 1893, p. 494.

Samar.

Tupaia javanica Horsf.

Tupaia javanica Horsf.; Everett, P. Z. S. 1893, p. 495.

Palawan, Calamianes.

Galeopithecus philippinensis Waterh.

Galeopithecus philippinensis Watetu.; Günther, P. Z. S. 1879, p. 74.

Basilan, Mindanao, Samar.

Mydaus marchei Heut.

Mydaus marchei Heut ; Everett, P. Z. S. 1893, p. 495.

Palawan, Calamianes.

Arctitis binturong Raffles.

Arctitis binturong Raffles; Everett, P. Z. S. 1893, p. 495.

Palawan.

Paradoxurus philippinensis Jourdan.

Paradoxurus philippinensis Jourd.; Günther, P. Z. S. 1879, p. 75.

Mindoro, Panay, Negros, Mindanao, Basilan, Palawan. Probably occurs on every island of any size in the group.

Viverra tangalunga Gray.

Viverra tangalunga Gray; Everett, P. Z. S. 1893, p. 495.

Siquijor, Panay, Mindoro, Mindanao, Palawan, Calamianes.
Like the preceding species, V. tangalunga ranges throughout the group.

Felis bengalensis Kerr.

Felis bengalensis Kerr; Blanford, P. Z. S. 1887, p. 631.
Felis minuta Temm.; Everett, P. Z. S. 1889, p. 223.

Panay, Negros, Cebu.

An animal which was in all probability a cross between this species and the common house cat was seen by us at the house of Mr. C. R. Blair Pickford, Toledo, Cebu.

Hystrix pumila Günther.

Hystrix pumila G.; Everett, P. Z. S. 1893, p. 495.

Palawan, Calamianes.

Sciuropterus nigripes Thos.

Sciuropterous nigripes Thos.; Everett, P. Z. S. 1893, p. 495.

Palawan.

Sciurus steerii Günther.

Sciurus steerii G.; Everett, P. Z. S. 1893, p. 496.

Palawan.

Sciurus samarensis Steere.

Sciurus samarensis Steere.   List B.& M. Steere Exped. p. 30.

Samar.

Sciurus mindanensis Steere.

Sciurus mindanensis Steere, List B. and M. Steere Exped. p. 29, 1890.

Mindanao, Basilan.

Sciurus mindanensis Steere and Sciurus cagsi Meyer seem to be synonyms.

Sciurus coccinus Thomas.

Sciurus coccinus Thomas; Ann. and Mag. Nat. Hist. (6) vol. ii, p. 407 (1888).

Sciurus philippinensis Steere, List B. and M. Steere Exped. p. 29.
Mindanao, Basilan.

Bubalus kerabau forus.

Bubalus kerabau ferus Nehring, Seitzungsbericht der Gesellschaft natur-forschender Freunde zu Berlin, No. 6, 1890, p. 101.
Bubalus indicus Steere, P. Z. S. 1889, p. 415.

Found wild by us in Mindoro, the Calamianes and Masbate. Oc-curs in the wild state in Luzon, Negros and Mindanao as well.

Bubalus mindorensis Heude.

Bubalus mindorensis Heude; Nehring, Sitzungsbericht der Ges. naturfor-schender Freunde zu Berlin, No. 6, p. 101, 1890.

Probubalus mindorensis Steere, List B. and M. Steere Exped. p. 28 (1890).

Mindoro.

Tragulus javanicus Gmel.

Tragulus javanicus Gmel.; Everett, P. Z. S. p. 223.

Balabac.

Manis javanica Desm.

Manis javanica Desm.; Everett, P. Z. S. 1893, p. 496.

Palawan, Calamianes.

Sus celebensis var. philippinensis.

Sus celebensis var. philippinensis Nehring. Tawi Tawi, Sulu, Basilan, Mindanao, Samar, Negros, Panay, Mindoro, Sibuyan, Luzon, Masbate.

Sus ahaenobarbus Heut.

Sus ahaenobarbus Heut.; Everett, P. Z. S. 1893, p. 496.

Palawan, Calamianes,

A considerable number of mammals, including deer from Masbate, Sulu, and the Calamianes islands, and bats from various localities, have not yet been identified. Skeletons of the mammals as well as skins were invariably collected while many of the small mammals as well as numerous birds were preserved entire in alcohol.

Those who wish to verify statements made in this paper will be given an opportunity to acquire the necessary material. Series of fifteen specimens showing the variations in color of *Ceyx bournsi* will be sent to any responsible person who will undertake to pay transportation charges and return them within a reasonable time.

Requests for such material may be addressed to Henry L. Osborn Hamline University, Saint Paul, Minnesota, or Dean C. Worcester. 9 Elm street, Ann Arbor, Michigan.

MINNEAPOLIS, December 8, 1894.

www.ingramcontent.com/pod-product-compliance
Lightning Source LLC
Chambersburg PA
CBHW022006190326
41519CB00010B/1412